Measurement of
Two Phase Flow Parameters

Measurement of Two Phase Flow Parameters

G. F. HEWITT

Engineering Sciences Division
AERE Harwell
Oxfordshire, U.K.

1978

ACADEMIC PRESS
London · New York · San Francisco

A Subsidiary of Harcourt Brace Jovanovich, Publishers

ACADEMIC PRESS INC. (LONDON) LTD.
24/28 Oval Road,
London NW1

United States Edition published by
ACADEMIC PRESS INC.
111 Fifth Avenue
New York, New York 10003

Copyright © 1978 by
G. F. HEWITT

All Rights Reserved

No part of this book may be reproduced in any form by photostat, microfilm, or any other means, without written permission from the publishers

Library of Congress Catalog Card Number: 78-72549
ISBN: 0-12-346260-6

Printed in Great Britain by
Whitstable Litho Ltd., Whitstable, Kent.

Preface

Two phase flow is not yet an area in which theoretical prediction of flow parameters is generally possible. Indeed, this situation is likely to persist for the forseeable future. Thus, the role of experiments and of parametric measurements is particularly important. This book is concerned with describing, first, the kinds of measurements which the design engineer and the research investigator may wish to make and then with listing the various techniques and evaluating their applicability. A vast range of such techniques is available and the present author would not claim that this text is comprehensive, though a large number of methods are described. The main purpose of the book is to classify the methods and to assist the research worker in approaching the design of experiments. However, the text is also aimed at being of general interest to those who use two-phase flow data and correlations. Since plant design in this area is heavily dependent on experimental data, it is important that the designer should appreciate the limitations of the measurements on which his design is based.

The present book is based partly on two previous reports. The first of these (by the present author) was presented at the Haifa two phase flow symposium in 1971 and was subsequently published in the Journal of the British Nuclear Energy Society and in Advances in Heat and Mass Transfer. Using the same basic structure, a much expanded report was produced in March 1976 by myself and my colleague, Mr P C Lovegrove. This latter work was done under contract to the Electric Power Research Institute, Palo Alto, California (Contract Number 446-1) and was published as EPRI Report No NP118. Though the present text has again been considerably expanded, I would like to acknowledge particularly the contribution Mr Lovegrove made in collecting and assessing material in this area. I would also like to thank his daughter, Miss Jenny Lovegrove, who provided invaluable clerical assistance. I would also like to acknowledge with thanks the support of EPRI and their permission to publish material contained in EPRI Report NP118.

The draft and final versions of this book were typed by my secretary, Mrs B Granito, and I would like to thank her for her patient, speedy and accurate work.

Preface

Finally, I would like to thank my many colleagues at Harwell for helpful advice and assistance in the preparation of this text.

G F Hewitt

Contents

Preface	v
1 Introduction to Two Phase Flow	1
Background	1
Applications of two phase flow	2
The physical nature of two phase flows	3
Analytical methods for steady two phase flows	7
Two phase heat transfer	13
Unstable and transient two phase flows	18
2 Classification of Two Phase Parameters	20
Introduction	20
Classification of parameters	21
Other sources of information	21
3 Primary Design Parameters	23
Introduction	23
Pressure drop	24
Heat transfer coefficient	30
Mass transfer coefficient	41
Mean phase content - void fraction	47
Critical heat flux	64
4 Primary Design Parameters (Fault Conditions)	69
Introduction	69
System discharge and related parameters	69
Bubble growth and collapse	76
Dryout under fault conditions	78
Rewetting	82
5 Secondary Design Parameters	85
Introduction	85
Vibration and transient momentum flux	85
Flow distribution	89
Stability	91
Quality and mass flow measurement	92
Liquid level detection	103
6 Second Order Parameters	105
Introduction	105
Flow pattern	105
Time averaged film thickness and amplitude distribution	111
Mass flow distribution	116

Contents

    Phase concentration distribution (local void fraction)   120
    Velocity and momentum flux distribution   128
    Concentration distribution   141
    Mixing characteristics of two phase flow (including phase mass transfer)   146
    Drop, bubble and particle size measurements   151
    Wall shear stress   161
    Temperature distribution   165
    Entrainment (film flow rate)   168
    Contact angle   173

7 Third Order Parameters   175
    Introduction   175
    Film thickness - interfacial waves   176
    Concentration of phases and components   180
    Velocity fluctuations   183
    Pressure fluctuations   185
    Temperature fluctuations   185
    Photographic observations of local phenomena   189
    Fluctuations in wall shear stress   194

8 Scaling of Two Phase Systems   196
    Introduction   196
    Application of fluorinated hydrocarbon scaling   196
    Development of scaling laws for use with fluorinated hydrocarbon scaling   204

References   209
Subject Index   275

1 Introduction to Two Phase Flow

Background

The objective of this introductory chapter is to provide a broad description of two phase flow phenomena with the aim of setting the scene for the discussion and description of measurement methods which form the body of this book.

Though methods for other multiphase systems are not specifically excluded from the book, it should be stated at the outset that the work has been concentrated mainly on gas-liquid systems. This partly reflects the author's own background, but it is also an indication of the relatively greater amount of work that has been done on gas-liquid flows.

The chapter begins by discussing the industrial applications of two phase flows and then goes on to discuss the nature of the flows (ie two phase flow patterns). Next, the various models available for multiphase flows are briefly surveyed starting with models which are unrelated to flow patterns as follows:
(a) One dimensional models in which only the relative velocity of the phases are considered.
(b) Two dimensional relative velocity methods.
(c) Homogeneous flow methods.
(d) Totally separated flow methods.

Models are then discussed which relate specifically to the flow patterns involved, and in particular the bubble flow, slug flow and annular flow regimes.

A brief description is then given of two phase heat transfer in both static ("pool boiling") and forced convective systems. Finally, the chapter deals briefly with unsteady flow situations - that is, with two phase flow instabilities and with transient phenomena.

As was stated above, this chapter is aimed at giving a broad general introduction and it is obviously impossible to go into detail in any given aspect. However, the reader may wish to read further about many of the aspects covered here and the text books of Tong (1965), Wallis (1969), Hewitt and Hall-Taylor (1970), Collier (1972) and Butterworth and Hewitt (1977) may be convenient sources.

Applications of Two Phase Flow

Two phase flows are ubiquitous in a whole range of industrial applications. For instance, a majority of all industrial heat exchangers involves two phase flows in one form or another. It is perhaps useful here to mention a few of the applications of the various forms of two phase flow:

Gas-liquid flows: Systems involving gas-liquid flows are found widely in the process, chemical, petroleum and related industries. Here, the problems exist on the transport of gas-liquid mixtures in pipelines and also in the design of equipment such as boilers, condensers, distillation towers, absorption towers. In these industries, the two phase mixtures are often multi-component and component mass transfer is often a limiting factor in design. Two phase flow is also important in the power industry in both conventional steam plant (where two-phase flow problems are encountered in the boiler, in the turbine and in the condenser systems) and in nuclear power systems. Indeed, it has been the necessity for better information relating to Boiling Water Reactors and Pressurised Water Reactors which has given the main impetus for the vast increase of interest and work in this area over the past three decades. Problems in nuclear reactors include not only steady state design (calculation of pressure drop, heat transfer coefficients, critical heat flux etc) but also problems of the system behaviour in transient, and particularly accident, conditions. Gas-liquid two phase flows are also important in industries such as sewage treatment, air conditioning and refrigeration and cryogenics.

Gas-solid systems: Pipeline flows of solids suspended in gases are important in combustion systems and in pneumatic conveying. Fluidised beds - which are becoming steadily more important, particularly for the combustion of solid fuels - may also be regarded as a form of gas-solids flow.

Liquid-liquid systems: These systems find application in the preparation and flow of emulsions and in mass transfer by liquid-liquid solvent extraction. In the latter context, extraction apparatus includes packed columns, pulsed columns, stirred contactors and pipeline contactors.

Liquid-solid flows: This type of flow is found in hydraulic conveying of solid materials and liquid-solid systems are of great importance in a wide variety of metallurgical extraction processes. Suspensions of solids in liquids also occur in crystallisation systems and in china clay extractions.

As will be seen from the above brief survey, two phase flows are not restricted only to channel flow systems, but also include a wide variety of alternative system geometries such as stirred vessels, the shell side of heat exchangers, packed beds etc etc. Very often, measurements are required on actual equipment involving two phase flow and the nature of the equipment obviously has a very direct bearing on the method chosen.

The Physical Nature of Two Phase Flows

For gas-liquid and liquid-liquid flows, the main complicating feature, distinguishing such flows from single phase gas or liquid flows, is the existence of deformable interfaces whose shape and distribution are of critical importance in determining the characteristics of the flow. In the case of gas-solids and liquid-solid flows, the interfaces are non-deformable but such flows are really complex since the distribution of the solid phase within the continuous fluid phase is generally unknown as are the details of the local interactions between the phases.

In channel flows, the specific interactions of the respective phases with the channel wall are of considerable importance in governing flow pattern and phase distribution. In gas-liquid and liquid-liquid flows, the presence of surface active agents can often critically affect the flow behaviour.

Fortunately, for flows with deformable interfaces, the effect of surface tension is to cause a tendancy towards the formation of curved interfaces which, for small bubbles and drops, leads to element shapes in the discontinuous phase which are approximately spherical. This, combined with other factors, leads to the possibility of a categorisation of the nature of the flows in to what are commonly called "flow patterns" or "flow regimes". Typical flow regimes for vertical gas-liquid flows are shown in Figure 1.1 and regimes for horizontal gas-liquid flows are illustrated in Figure 1.2. The regimes in

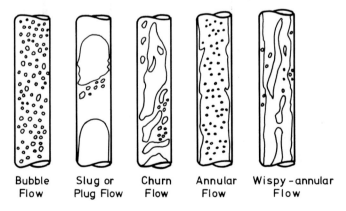

Figure 1.1: Two phase gas-liquid flow regimes in vertical upwards flow.

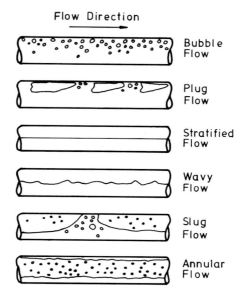

Figure 1.2: Two phase gas-liquid flow regimes in horizontal flow.

vertical flow are as follows:
(1) *Bubble flow*. Gas bubbles flow in a continuum of liquid.
(2) *Slug or plug flow*. When the concentration of bubbles in bubble flow becomes high, more coalescence occurs and, progressively, the bubble diameter approaches that of the tube. The large, bullet-shaped bubbles characteristic of slug or plug flow (the names are used interchangeably in vertical flows) are then encountered.
(3) *Churn flow*. With further increases of gas flow rate, the velocity of the slug flow bubbles increases and, ultimately, a breakdown of these bubbles occurs leading to an unstable regime in which there is often an oscillatory motion of the liquid in the tube.
(4) *Annular flow*. In this case, the liquid flows on the wall of the tube as a film and the gas flows in the centre. Usually, some of the liquid phase is entrained as small droplets in the gas core.
(5) *Wispy annular flow*. As the liquid flow rate is increased, the droplet concentration in the gas core of annular flow increases and, ultimately, droplet coalescence occurs leading to large lumps or streaks (or wisps) of liquid occuring in the gas core. This regime is characteristic of high mass velocity flows.

For horizontal flows, the main complicating feature is that gravitational forces act on the liquid phase causing it to be displaced towards the bottom of the channel. The regimes shown in Figure 1.2 are defined as follows:
(1) *Bubble flow*. In horizontal tubes, the bubbles tend to

flow at the top of the tube as illustrated.
(2) *Plug flow*. Again, the characteristic bullet-shaped bubbles occur, but they tend to move along in a position closer to the top of the tube.
(3) *Stratified flow*. Here, the gravitational separation is complete; liquid flows along the bottom of the tube and gas along the top of the tube.
(4) *Wavy flow*. As the gas velocity is increased in stratified flow, waves are formed on the gas-liquid interface giving the "wavy flow" regime.
(5) *Slug flow*. In this regime, large frothy slugs are formed (for instance by growth of the waves in wavy flow) and are transported rapidly along the channel. In some cases, these slugs occupy the whole of the cross section of the channel and, in other cases, they are in the form of a very large surge wave on the thick liquid film at the bottom of the channel. In both cases, liquid is deposited at the top of the channel and gradually drains downwards between the slugs.
(6) *Annular flow*. In horizontal tubes, annular flow occurs at high gas flow rates. As in vertical flows, there is usually some entrainment of the liquid phase as droplets in the gas core and, as a result of the gravitational effects, the film at the bottom of the tube is often much thicker than the film at the top.

A great deal of effort has gone into investigating and classifying regimes occuring under various flow conditions and the usual outcome is to express results in terms of a "flow pattern map". For illustration purposes, two such maps are given here as follows:
(1) Figure 1.3 shows the map due to Hewitt and Roberts (1969) for vertical flows. In this map, the various flow patterns are plotted in terms of the superficial momentum fluxes (product of the square of the superficial velocity and the phase density) of the respective phases.
(2) Figure 1.4 shows the map of Mandhane et al (1974) for horizontal flows. Here the map is plotted in terms of the superficial velocities of the respective phases.

Obviously, representation of two phase flow patterns in terms of generalised maps must be the subject of considerable empiricism and there has been a continuing search for a more fundamental understanding of the nature of flow pattern transitions. A discussion of the work in this area is beyond the scope of this present chapter and the reader is referred to the review by Hewitt (1976). The flow patterns in liquid-liquid flows are of a somewhat similar nature to those described above for gas-liquid flows. In gas-solids and liquid-solid flows, a distinction can be made between "dispersed" and "fluidised" systems. In fluidised beds (where there is no net motion of the solid phases, though the solid does circulate within the

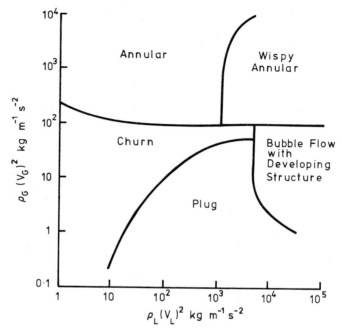

Figure 1.3: Flow pattern map of Hewitt and Roberts (1969) for vertical upwards flow.

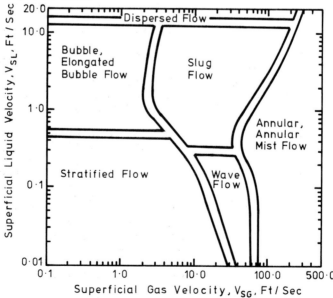

Figure 1.4: Flow pattern map of Mandhane et al (1974) for horizontal gas-liquid flow.

bed) the occurrence of large bubbles somewhat similar to those found in slug flow in gas-liquid systems is of particular importance in governing the behaviour of the beds.

Analytical Methods for Steady Two-Phase Flows

One-Dimensional (Constant Relative Velocity) Methods

This class of methods is aimed specifically at the calculation of the fractions of the channel volume (or channel cross sectional area) occupied by the respective phases. An example of such a calculation would be of the hold-up of solid particles in a gas-solids flow or the "void fraction" (fraction of channel cross section occupied by the gas phase) in a gas-liquid bubble flow. The methods do not lead directly to the calculation of friction at the wall and, indeed, the wall shear stress is ignored in the calculations. It is also assumed that there is no variation in phase content or phase velocities across the channel, hence the term "one dimensional". Only a brief survey of the method is given here; the reader is referred to the textbooks of Wallis (1969) and/or Butterworth and Hewitt (1977) for a more detailed treatment.

Suppose that one requires to calculate the fraction of the channel cross section occupied by phase 1 in a mixture of phases 1 and 2 flowing at superficial velocities V_1 and V_2 respectively. Here, the superficial velocity is defined as the total volumetric flow rate of the phase divided by the channel cross-sectional area. It is convenient to consider relative phase motion through a plane which is travelling at the total superficial velocity V ($V=V_1+V_2$). Consider, for example, a vertical upwards flow: by continuity, the flux of the lighter phase through the plane moving upwards at velocity V must be equal and opposite to the flux of the heavier phase through this plane. From continuity considerations, it can be shown that the flux of, say, component 1 through the plane moving at velocity V is given by:

$$j_{12} = V_1(1-\alpha_1) - \alpha_1 V_2 \qquad (1.1)$$

where α_1 is the fraction of the cross section of the channel occupied by phase 1. In addition to this continuity relationship, it can be postulated that j_{12} can also be obtained from an expression as follows:

$$j_{12} = \text{fn}(\alpha_1, \text{physical properties, flow pattern}) \qquad (1.2)$$

which is an expression of the physics of the system. For

instance, if the flow pattern is bubbly flow, relative motion of the phases will be by bubbles rising through the liquid with a rise velocity which is a function of the physical properties. In that case, for low values of α_1 where there is little interference between bubbles, j_{12} would simply be the product of α_1 and the bubble rise velocity.

In order to obtain the phase content, equations 1.1 and 1.2 must be solved simultaneously. An example of this is given below for gas-liquid bubble flow.

Two-Dimensional Relative Velocity Methods

The one-dimensional relative velocity method assumes constant velocity and phase content across the channel. In practice, this is rarely true. For the case where conditions vary across the channel, Zuber and Findlay (1965) suggest the use of "average" values defined as follows:

$$<F> = \frac{1}{A} \int_A F \, dA \tag{1.3}$$

and "weighted mean" values defined by:

$$\bar{F} = \frac{<\alpha F>}{<\alpha>} = \frac{\frac{1}{A}\int_A \alpha F \, dA}{\frac{1}{A}\int_A \alpha \, dA} \tag{1.4}$$

where F is a flow parameter (eg gas velocity, liquid velocity etc) and A is the cross sectional area of the channel. The local velocity, u_1 of phase 1 can be expressed as:

$$u_1 = V + u_{1V} \tag{1.5}$$

where V is the total local superficial velocity and u_{1V} is the velocity of phase 1 relative to a point moving at velocity V. The weighted mean value of the phase 1 velocity is given by:

$$\bar{u}_1 = C_o <V> + \frac{<\alpha_1 u_{1V}>}{<\alpha_1>} \tag{1.6}$$

where C_o is given by:

$$C_o = \frac{<\alpha V>}{<\alpha><V>} \tag{1.7}$$

Another parameter often referred to in two phase flow studies is the so-called "slip ratio" S_{12} which is defined as the ratio between $\bar{u}_1$ and $\bar{u}_2$ and is given in terms of the above averages by the expression:

$$S_{12} = \frac{\bar{u}_1}{\bar{u}_2} = \frac{1 - <\alpha>}{\dfrac{1}{C_o + \dfrac{<u_{1V}\alpha_1>}{<\alpha><V>}} - <\alpha>} \quad (1.8)$$

For the limiting case where there is no <u>local</u> relative motion between the phases ($u_{1V} = 0$) equation 1.8 reduces to:

$$S_{12} = \frac{1 - <\alpha>}{(1/C_o) - <\alpha>} \quad (1.9)$$

Thus, even in the case where there is no <u>local</u> relative motion between the phases, the overall slip ratio between the phases can have a value different from unity due to the weighting of the profiles of phase content and superficial velocity. This was first pointed out by Bankoff (1960).

Two Phase Flow Momentum Balance and the Homogeneous Model

Calculation of pressure drop is important in two phase flow systems and a wide variety of models have been proposed. The total pressure gradient (dp/dz) can be considered to be composed of three components arising respectively from friction losses (dp_F/dz), acceleration of the fluid (dp_a/dz) and change of pressure due to gravitational forces (dp_g/dz). A momentum balance on the element of channel leads to the following expressions for the various terms:

$$-\frac{dp}{dz} = -\frac{dp_F}{dz} - \frac{dp_a}{dz} - \frac{dp_g}{dz} \quad (1.10)$$

$$-\frac{dp}{dz} = \frac{1}{A}\int_S \tau_o \, dS + \frac{1}{A}\int_A \frac{d}{dz}[G_1 u_1 + G_2 u_2]\, dA$$
$$+ \frac{1}{A}\int_A g\sin\theta\,[\alpha_1 \rho_1 + (1-\alpha_1)\rho_2]\, dA \quad (1.11)$$

Where S is the tube periphery, τ_o the wall shear stress, z the axial distance along the channel, θ the angle of inclination of the tube from the horizontal, G_1 and G_2 local mass fluxes and u_1 and u_2 the local velocities of the respective phases and ρ_1 and ρ_2 the respective phase densities. Further details of the derivation (and of the derivations of the equations below) are given, for instance, by Hewitt and Hall-Taylor (1970).

It will be noted that information on local mass fluxes and velocities and local phase contents are required for the calculation of the accelerational and gravitational components in equation 1.11. This requirement partly explains the emphasis placed on phase content measurement in two phase flows.

The normal procedure for using equation 1.11 is to make some (arbitrary) assumption about the nature of the flow. One such assumption would be to treat the flow as a homogeneous mixture of the phases with constant phase content and constant (and equal) velocities for the two phases across the whole channel. This is the so-called "homogeneous model" for two phase flow. Inserting the further assumption of constant wall shear stress, the following version of equation 1.11 is obtained:

$$-\frac{dp}{dz} = \frac{S\tau_o}{A} + \frac{G^2 d(1/\rho_H)}{dz} + g\rho_H \sin\theta \qquad (1.12)$$

where G is the mean total mass flux and ρ_H is the "homogeneous mean density" given by:

$$\frac{1}{\rho_H} = \frac{x_1}{\rho_1} + \frac{(1-x_1)}{\rho_2} \qquad (1.13)$$

where x_1 is the "quality" for phase 1 - ie the ratio of the mass rate of flow of phase 1 to the total mass rate of flow. The accelerational and gravitational terms are readily calculated and the frictional term is often estimated by analogy with single phase flow by introducing a two phase flow friction factor f_{TP} as follows:

$$-\frac{dp_F}{dz} = \frac{S\tau_o}{A} = \frac{1}{2}\frac{S}{A}\frac{f_{TP}G^2}{\rho_H} \qquad (1.14)$$

f_{TP} can either be correlated directly or can be estimated from single phase flow charts in terms of a two phase Reynolds number. Usually, the homogeneous model for calculation of pressure drop is not too satisfactory.

Separated Flow Models

An alternative to the homogeneous model is to consider the flow to be completely separated within the channel into two zones, one occupied by phase 1 and the other by phase 2. For this case, and again making the assumption of constant wall shear stress around the periphery of the channel, equation 1.11 reduces to:

$$-\frac{dp}{dz} = \frac{S\tau_o}{A} + G^2 \frac{d}{dz}\left[\frac{x_1^2}{\alpha_1 \rho_1} + \frac{(1-x_1)^2}{(1-\alpha_1)\rho_2}\right] + g\sin\theta[\alpha_1 \rho_1 + (1-\alpha_1)\rho_2]$$

(1.15)

In the separated flow models, a knowledge of phase content is essential to calculate the accelerational and gravitational terms. The most common way of correlating the frictional term is to express it in terms of a "friction multiplier" ϕ^2_1 or ϕ^2_2 which is defined as the ratio of the frictional pressure gradient to that for one or other phase flowing alone in the channel. Thus, for example:

$$\phi_1 = \left[\frac{(dp_F/dz)}{(dp_F/dz)_1}\right]^{\frac{1}{2}}$$

(1.16)

ϕ can be correlated in terms of a number of parameters; perhaps the best known correlation is that of Lockhart and Martinelli (1949) who correlated ϕ for gas-liquid flow in terms of the parameter X defined as follows:

$$X = \frac{(dp_F/dz)_1}{(dp_F/dz)_2}$$

(1.17)

Where, for gas-liquid flow, phase 2 is taken as the gas phase. Many developments of the Lockhart and Martinelli method have been reported, noteable amongst which are those of Baroczy (1965), Chisholm (1967) and Chisholm and Sutherland (1969). A more detailed review is given in the texts by Collier (1972), Hewitt and Hall-Taylor (1970) and Butterworth and Hewitt (1977).

Models Based on Flow Patterns

The homogeneous and separated flow models tend to give an inadequate representation of real two phase flows and the detailed physics of the flow (including the flow pattern) are important. Here, for illustration, one may consider methods which are specifically related to flow patterns in gas liquid flow.

Bubble flow. Zuber and Findlay (1965) suggested semi-empirical relationships for j_{12} of the form:

$$j_{12} = u_\infty \, \alpha_1 (1-\alpha_1)^n \qquad (1.18)$$

which can be solved simultaneously with the continuity relationships (equation 1.1). In equation 1.18, u_∞ is the rise velocity of a single bubble in an infinite liquid pool and n is an exponent which varies with the form of the bubble flow. More details are given in the original reference and also by Butterworth and Hewitt (1977).

Slug or plug flow. Here, nearly all of the gas may travel in the large slug flow bubbles and therefore has a well specified velocity - ie that of the slug flow bubble, u_S. If the average superficial velocity of the gas (phase 1) entering the channel is V_1 then the average void fraction is given simply by $\alpha = V_1/u_S$. For low viscosity fluids in tubes which are not of small diameter (eg water in a 20mm bore tube) then the value of u_S can be calculated from:

$$u_S = 0.345\sqrt{gd_o} \qquad (1.19)$$

where g is the acceleration due to gravity and d_o the diameter of the tube. For small diameter tubes, surface tension effects become important. Also, the effect of liquid viscosity can become significant. Modified expressions taking account of these effects are discussed, for instance, by Wallis (1962).

Annular flow. In annular flow in particular, the state of development of the flow is critical. Equilibrium flow conditions are reached very slowly indeed and, for most practical applications, the flow is non-equilibrium. Non-equilibrium flow models can be built up using a number of basic relationships of the following types:
(1) A "triangular relationship" which allows the calculation of liquid film thickness, say, from a knowledge of the flow rate in the film and the pressure gradient.
(2) An interfacial roughness relationship. The wavy interface in annular flow acts, in effect, as a surface roughness which increases the pressure drop of the gas core flow. It is found, over limited ranges, that the interfacial roughness depends only on the thickness of the liquid film.
(3) Relationships for the entrainment and deposition of

liquid droplets from and to the film.
The above relationships can be integrated from known (or arbitrarily chosen) boundary conditions giving a prediction of film thickness (and hence liquid phase content), film flow rate and pressure drop along the channel. The basis of these methods are surveyed in the texts by Hewitt and Hall-Taylor (1970) and Butterworth and Hewitt (1977). Details of applications to evaporating flows (and specifically to the prediction of dryout) are reported, for instance, by Whalley et al (1974) and by Hewitt (1976).

Two Phase Heat Transfer

In many practical applications of two phase flows, heat transfer is taking place between the phases and within the phases. Heat transfer problems are important in all forms of two phase flow, but for the purposes of this chapter, the following brief review is restricted to gas-liquid systems. This is because it is felt particularly important for the general reader to have an insight into some of the phenomena occurring here.

It is useful first to consider the question of thermodynamic equilibrium in the various states of a single component fluid. The relationship between pressure and specific volume for a given temperature below the critical temperature is illustrated in Figure 1.5 (after W B Hall, University of Manchester). Along the lines AB and EF the substance exists as a single-phase liquid and as a single vapour respectively. When the pressure falls to p_{SAT} along AB or rises to p_{SAT} along FE, a condition is reached in which the liquid or vapour can be in stable equilibrium as a two phase system. In this equilibrium state, the phase

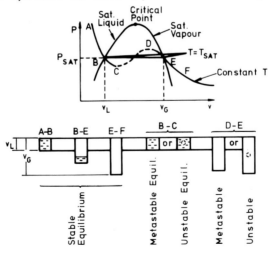

Figure 1.5: Relationship between pressure and volume for a pure substance.

specific volumes are for "saturated" vapour and liquid (namely v_L and v_G respectively). The specific volume of an equilibrium vapour-liquid mixture can have any value between v_L and v_G (ie along the line BE) depending on the content. In practice, it is possible to reduce liquid pressure or increase the vapour pressure beyond p_{SAT} without necessarily inducing a phase change. For these latter cases, the system follows line BC or ED respectively and is in metastable (single phase) or unstable (spherical bubble or droplet) equilibrium as illustrated. Along the line BE, the vapour and liquid are in equilibrium when separated by a planar interface. Along the lines BC and DE, equilibrium can only be maintained if the interface is curved giving a change of pressure from liquid to vapour. It will be seen from Figure 1.5 that deviations from equilibrium can occur even in adiabatic two phase systems. In heated systems, and in transient situations, other forms of departure from equilibrium are observed. Broadly speaking, departures from equilibrium can be classified as follows:
(1) "Continuum effects" (as illustrated in Figure 1-5):
   (a) It is possible for vapour to be cooled below the saturation temperature without formation of the liquid phase (in the form of liquid droplets).
   (b) It is possible for a liquid to be superheated without formation of the vapour phase.
In both of the above cases, the problem is that the discontinuous phase has to be nucleated within the continuous phase. Actually, quite large departures from equilibrium are required for such nucleation within the continuous phase itself. More commonly, the first formation of the discontinuous phase occurs at the vessel or channel wall enclosing it (this is called "heterogeneous" nucleation).
(2) "Transient effects":
   (a) Droplets in superheated vapour often evaporate relatively slowly and their presence at significant superheats represents a departure from equilibrium.
   (b) Vapour bubbles can exist similarly in sub-cooled liquid and can take a considerable time to condense particularly if the sub-cooling is small.
(3) "Local" effects:
   (a) If the channel wall is being heated, then steam bubbles can be formed at the wall ("sub-cooled boiling") even though the bulk mixed enthalpy of the fluid is well below that for saturation.
   (b) If the channel wall is being cooled, with a superheated vapour flow in the channel, then liquid can be formed at the channel wall even though, again, the bulk enthalpy of the vapour is well above that for saturation.
Systems which are not in thermodynamic equilibrium present some of the most difficult challenges in measurement. The unambiguous determination of phase contents and tempera-

Two Phase Heat Transfer

tures in such systems has yet to be fully achieved.

It is beyond the scope of this present book to go in to the details of phenomena occurring during two phase heat transfer. However, by way of illustration, two examples of heat transfer during evaporation will be described. The first is evaporation from a horizontal tube or wire suspended in a static pool of liquid ("pool boiling") and the second is that of evaporation liquid during forced convective flow through a tube.

The sequence of events which occur during pool boiling is illustrated in Figure 1.6. As the temperature of the heater is increased, heat is transferred by conduction and by convection currents (a) and, as the heater's surface temperature is increased further, the first bubbles are seen on the surface of the heater but do not detach. With further increases of heat flux, the bubbles increase in number and some detach to condense in the liquid. In this region, the extra convection etc induced by the boiling so improves the efficiency of heat transfer that the amount of heat transferred rises rapidly with only small increases of surface temperature. However, a point is reached at which the bubbles at the heated surface coalesce to give partial vapour blanketing (b). This leads to a rapid decrease in the amount of heat transferred since the vapour blanket forms a major resistance to the passage of heat. The heat flux corresponding to the transition from the nucleate boiling to the vapour

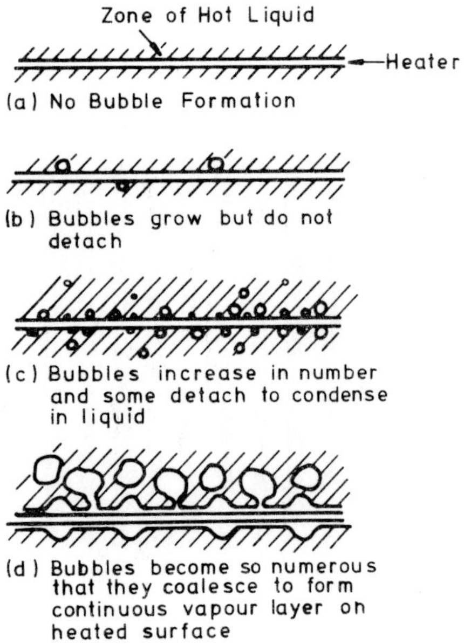

(a) No Bubble Formation

(b) Bubbles grow but do not detach

(c) Bubbles increase in number and some detach to condense in liquid

(d) Bubbles become so numerous that they coalesce to form continuous vapour layer on heated surface

Figure 1.6: Stages in sub-cooled boiling from a horizontal heated element.

blanketing (film boiling) region is often referred to as the "critical heat flux"; if the heater is powered electrically then the reduction in heat transfer efficiency that occurs after the critical point results in a rapid overheating of the surface, sometimes destroying it. The critical phenomenon is also sometimes called "burnout", "boiling crisis", "dryout" and "DNB" (departure from nucleate boiling).

Studies of pool boiling have strongly coloured the interpretation of boiling phenomena in channel flow. However, the nature of the phenomena, particularly that of the critical heat flux, can be very different in channel flow as is illustrated in Figure 1.7. In this conceptual diagram, liquid at constant velocity and temperature enters a tube and the state of flow and heat transfer in the tube for equal successive steps in heat input is illustrated in Figures 1.7A-1.7U. The increment in heat flux is equal to that required to heat the input liquid to the saturation temperature at the end of the tube. Note that the saturation temperature corresponds to a vapour quality (ratio of vapour flow to total flow - usually calculated on a thermodynamic equilibrium basis) of zero. The flow patterns develop from single phase liquid flow through bubble flow, slug flow, churn flow and into annular flow. The first generation of vapour takes place by nucleation at the wall and the locus of the onset of nucleation is line XX. If the heat flux is low (Figure 1.7B) then nucleation may be delayed beyond the point at which the thermodynamic quality, x, is 0. In this case the wall temperature is insufficient to cause

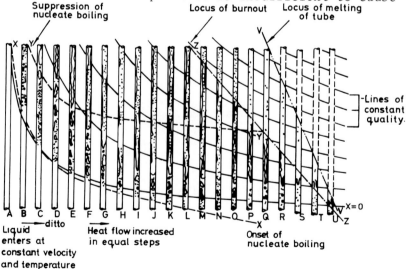

Figure 1.7: Regimes of fluid flow and heat transfer in the evaporation of liquid in a tube.

nucleation and the liquid can have a considerable bulk superheat before any vapour is formed. At high heat fluxes, however, nucleation occurs upstream of a locus of zero thermodynamic quality due to sub-cooled boiling. Initially, in the sub-cooled boiling region, the vapour bubbles tend to remain near the wall. However, as the heat flux is further increased, the bubbles detach and undergo condensation to an extent depending on the liquid bulk temperature in the core of the flow. As the fluid proceeds up the tube, further generation of vapour takes place at the nucleation centres and by direct evaporation from the interfaces. The temperature driving force necessary to transfer the heat through the liquid phase to the interface by conduction and convection, decreases as the quality increases. This means that interface heat transfer increases in importance with respect to nucleate boiling heat transfer and, as quality increases, nucleate boiling is suppressed since the wall temperature is no longer high enough to maintain active nucleation sites. Complete suppression is indicated in Figure 1.7 by the locus YY. In the annular flow regime, liquid is lost from the liquid film by evaporation and by entrainment. Eventually, the liquid film flowrate at the end of the channel is reduced to zero (Figure 1.7L) and the wall becomes dry. It is this form of critical condition, rather than the film boiling transition illustrated in Figure 1.7, that is most common in forced convective boiling. For constant heat input, the temperature in the dry region is considerably above that in the region below the dryout point. On further increase of the heat flux, the dryout point is propagated downstream as shown by the locus ZZ. In most experiments on burnout, the test is terminated by cutting off the heat flux as soon as the first temperature excursion at the end of the channel occurs. If the heat flux is high enough, the temperature rise at burnout can cause melting of the channel and possible locus of this is illustrated by line VV in Figure 1.7. It should be emphasised that the position of VV with respect to Z can vary considerably with mass velocity and, in common with the other locii given in Figure 1.7, the actual position shown should be regarded as being for illustration only. The situations illustrated in Figure 1.7R-1.7U are physically impossible due to the melting of the tube and, to achieve burnout under the heat flux and inlet flow conditions, shorter tubes must be used. It will be noted that the line ZZ crosses the lines of constant quality and the regime of flow in which burnout occurs changes from the annular through to the sub-cooled boiling region. In this latter region, the mechanism is most probably connected with the onset of film boiling (of Figure 1.6 for pool boiling) rather than with the disappearance of a liquid film as in the annular regime.

Unstable and Transient Two Phase Flows

Two phase flow in many industrial and power systems is subject to a variety of excursive and oscillatory instabilities. Amongst the types of instability which can occur are:
(1) Excursive or "Ledinegg". Here, for given system heat input, a plot of pressure drop versus mass velocity passes through a maximum and a minimum. Thus, for a given pressure drop, several values of mass flow are possible and the system can go through an excursion from one condition to another.
(2) Chugging instability. This is caused by the liquid in the system becoming superheated because of the absence of suitable nucleation centres on the tube wall. Eventually, when nucleation finally does take place, a rapid vapour growth occurs forcing the liquid out of the channel and leading to a drop in temperature. The nucleation centre is suppressed and the whole cycle is repeated.
(3) Oscillatory instability. In steady flow, the various components of pressure drop mentioned in section 4 above are all additive. However, if the inlet flow to the channel is oscillating, then there are phase lags between the oscillations in the various components of pressure drop and, at a given frequency, the net effect on pressure drop is zero. In other words, oscillations can occur in inlet velocity without any compensating increase in pressure drop. This is perhaps the most important form of instability.

Even more important than problems of instability in two phase flow systems are problems connected with transient flows. In particular, the great interest in safety problems for boiling water and pressurised water reactors in recent years has let to extensive analytical and experimental studies of transients. Here, the main problem is trying to predict what would happen in the very unlikely event of, say, a break in one of the main circulation pipes of the reactor. Here, the problem is one of predicting what happens immediately after the break when the reactor fluid passes rapidly out through the break ("blow down") and when emergency cooling fluid is introduced into the reactor. Some of the main points of interest here are as follows:
(1) Ejection of fluid through the break is limited by the critical discharge rate. Critical flow in two phase systems is highly complex; a particularly important factor is the thermodynamic state of the fluid passing through the break. So rapid is the flow through the break that there is insufficient time for the attainment of thermodynamic equilibrium.
(2) Prediction methods for transients are being continuously improved. The most common method is to use numerical

predictions of system performance, splitting the total system into a series of elements (or "nodes"). The most common approach is to treat the flow in the transients as homogeneous though more recent models include simple representations of slip. Recently, it has been possible to increase the sophistication of transient modelling; for instance, the annular flow model described in section 4.5 above has been applied to transient flow and heat transfer in a reactor fuel element with encouraging results.

(3) The two phase flow and heat transfer mechanisms during reflooding of the reactor with emergency cooling water are very complex. The fuel element surfaces are often quite hot when the cooling water is introduced and there is a delay in the rewetting of the surfaces. This is mainly due to the fact that rapid axial conduction along the fuel element wall from the hot zone to the cooled zone leads to intense nucleate boiling and inhibition of wetting at the interface between the two zones.

Obviously, in a short chapter like this present one, there is little scope for detailed discussion of the problems of nuclear reactors; the reader is referred to the monograph by Lahey and Moody (1977) for a more detailed exposition.

Clearly, making measurements in unsteady and transient flows presents a considerable challenge to the experimentalist. In recent years, a vast expenditure of effort and money has been made in this area and many new instrumentation systems have been developed. Nevertheless, there is still much more development work to be done to achieve, in particular, reliable and robust techniques for transient conditions.

2 Classification of Two Phase Parameters

Introduction

A large number of alternative approaches could have been made to the classification of experimental methods. The two most obvious ones are:
(a) Classification according to the type of method (local or average, electrical or optical, interfering or non-interfering etc).
(b) Classification in terms of the parameters to be measured.

Both of these systems have merits but, the second has been chosen for this book for the following main reasons:
(a) The main aim of the book is to be of practical assistance to the experimentalist. Direct reference to the particular parameter, and the alternative methods of measuring it, seems preferable in this context.
(b) An important secondary aim of the book is to give the designer a feel for the way in which parameters of interest to him have been measured. Again, classification by parameter seems preferable.

Classification of Parameters

The following classification of parameters has been adopted:
(1) First order parameters. These are of direct relevance in design and, for the following presentation, have been further divided into three categories as follows:
  (a) Primary design parameters (steady state) (Chapter 3). These include parameters such as pressure drop and maximum heat-flux, which represent limits on the design.
  (b) Primary design parameters (fault conditions) (Chapter 4). In nuclear reactor systems, nowadays, the designer must be concerned, at a very early stage, with postulating fault conditions and designing a system to overcome their effects. In order to do this, the designer needs information on such things as discharge rates, the detection of faults and on methods of ensuring the rewetting of fuel elements which have dried out and become

overheated.
- (c) Secondary design parameters (Chapter 5). These are parameters which need to be investigated as part of the overall design but would not necessarily be the first consideration of the designer. Here, the discussion covers vibration problems, flow distribution problems and flow or quality monitoring.
(2) Second order parameters (Chapter 6). Here, one is concerned with parameters which are of interest in research on the system, aimed at obtaining better designs. In order to obtain a fuller understanding, leading to better design correlations for the first order parameters, it is often necessary to measure such things as two-phase flow pattern, mass-flow distributions, wall shear-stress and liquid entrainment. The designer is not going to calculate these quantities in designing a system, but he may nevertheless be interested in them, in order to get a better qualitative understanding of what is going on. The essential nature of second order parameters is that they are measured as part of the investigation of a system and are time-averaged or steady-state values.
(3) Third order parameters (Chapter 7). Two-phase flows are characterised by a fundamentally unsteady state. An understanding of the fluctuations of such quantities as velocity, phase content, temperature and wall shear-stress can aid in the development of improved design methods, and, hence, better systems. Such fluctuating parameters are referred to here as "third order". Again, the designer is unlikely to use this information directly in this work but it still may be of interest to him. Included in this area is the use of high-speed photographic techniques for the observation of fluctuating local phenomena and it is through the use of advanced optical and photographic techniques that considerable progress in the understanding of two-phase flows has been achieved.

Naturally, having set up a classification of the various areas of work, the division of the various topics must be, to some extent, arbitrary and the author has had some difficulties in categorising all the papers considered. However, it is believed that the above classification does give a suitable and relevant framework in which to consider the various experimental techniques and equipment.

In recent years, modelling of two-phase flow systems, particularly using fluorinated hydrocarbons as substitutes for high pressure water, has become very popular. This, in itself, can be regarded as an experimental technique and a brief review is given in Chapter 8.

Other Sources of Information

The importance of two-phase instrumentation has led to the generation of a number of other reviews to which the

reader may wish to refer. These include the following:
(1) A series of reviews produced at CENG, Grenoble. These are published in Delhaye (1972), Delhaye et al (1973a), Delhaye (1974) and Delhaye and Jones (1975). These surveys are particularly useful in considering statistical measurements.
(2) A review on electrical probes for two-phase flow measurements has been presented by Bergles (1969).
(3) A good source of information on Russian and other work, and particularly in the context of the use of the electro-chemical technique for measuring wall shear-stress and its characteristics in two-phase flow, is the book by Kutateladze (1973). Some aspects of this work are reviewed in what follows, but the book gives a consistent presentation.
(4) Problems of metering and instrumentation in gas-solid flows are reviewed by Boothroyd (1971).
(5) A review of instrumentation accuracy for a semiscale blowdown rig is given by Feldman and Naff (1975).
(6) A review of single and two phase flow measurement techniques for reactor safety studies produced by Brockett and Johnson (1976). This is a particularly useful source for selection of measuring instruments including, in particular, pressure transducers and temperature sensors.
(7) The book "Measurement techniques in heat transfer", edited by Eckert and Goldstein (1970). This is a very useful source on the various techniques for temperature measurement and also includes information on measurements of thermophysical properties and local velocity.

In addition to the above, a good general text on the fundamentals of temperature, pressure and flow measurement is that of Benedict (1969).

Most measured quantities in two phase flow are some kind of space and/or time average. Great care has to be exercised on the precise interpretation of the type of spatial (line, area, volume) average which has been taken. It is particularly important to be aware of the averaging process in any use of the quantities in comparing experiment and prediction. A detailed discussion of this problem is given by Delhaye and Achard (1977).

3 Primary Design Parameters

Introduction

The following parameters are covered in this chapter:
(a) *Pressure drop*: This is of obvious interest since it governs the pumping power required in the system. Pressure drop (as indicated in Chapter 1) is composed of frictional, gravitational and accelerational components but these are not separated in measurement.
(b) *Heat transfer coefficient*: In two phase heat transfer the coefficient between the surface and the two phase stream is often so high that the overall rate of heat transfer is governed primarily by other parts of the system - eg by the cooling water heat transfer in a condenser or by the radiant heat transfer in a boiler. However, there are many cases where the two phase heat transfer coefficient is of prime importance. The most significant class of applications here are those in which a fluid is being evaporated from a surface, the other side of which is heated by a condensing fluid. Examples here would be steam heated kettle or thermosyphon reboilers on a process plant or the matrix oxygen reboiler in a tonnage oxygen plant.
(c) *Mass transfer coefficient*: Inter-phase mass transfer coefficients are of significance in the design of mass transfer equipment such as fractionation columns and liquid-liquid extraction systems. Also, mass transfer effects may provide a limitation in some heat transfer processes (the prime example here is that of condensation in the presence of inert gases). Measurement of mass transfer coefficients can also assume importance in modelling heat transfer system through the analogy between heat transfer and mass transfer.
(d) *Mean phase content - void fraction*: Phase content can be important for a variety of reasons:
   (1) If valuable materials are being processed, the quantity (holdup) of such materials in the system is often economically significant.
   (2) In the design of boiling water nuclear reactors, estimation of the neutron absorption by the liquid

phase is of direct importance and, hence, void fraction is a primary design parameter.

(3) As was shown in Chapter 1, the gravitational and accelerational components of pressure drop are both dependent on phase content (void fraction).

(e) *Critical heat flux*: Although heat transfer coefficients are generally high in boiling systems, if the flux exceeds the critical value then a usually dramatic drop occurs in the coefficient. It is important, therefore, in design to have a knowledge of the critical flux.

Pressure Drop

Pressure drop measurement in two phase flow systems is usually achieved using either manometric or pressure transducer techniques.

Figure 3.1 illustrates schematically pressure drop measurement using a manometer. A pressure balance can be carried at level A as follows:

$$p_1 + (z_2-z_1)g\rho_c = p_2 + (z_4-z_3)g\rho_c + (z_3-z_1)g\rho_m \quad (3.1)$$

and re-arranging we have:

$$p_1 - p_2 = (z_3-z_1)g(\rho_m-\rho_c) + (z_4-z_2)g\rho_c \quad (3.2)$$

If $p_1 = p_2$ then the manometric difference is given by:

$$(z_3-z_1) = -(z_4-z_2)\frac{\rho_c}{\rho_m-\rho_c} \quad (3.3)$$

Thus, there is an "offset" on the manometer which depends on the distance between the tappings and the denisty (ρ_c) in the lines. In the absence of flow through the tube it follows that:

$$p_1 - p_2 = g\rho_t(z_4-z_2) \quad (3.4)$$

and the manometric difference is:

$$(z_3-z_1) = \frac{\rho_t-\rho_c}{\rho_m-\rho_c} \; z_4 - z_3 \quad (3.5)$$

The manometer will have zero differential if the fluid in

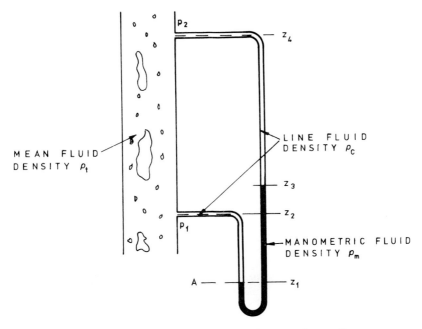

Figure 3.1: Measurement of pressure drop in two phase flow systems.

the line has the same density as the fluid in the tube.

It will be seen from the above derivation that it is of vital importance to know the composition and density of the fluid within the connection lines and, in practice, this means that it is important to control the lines in such a way that they are filled with single phase fluid corresponding to either the gas or the liquid phase flowing in the pipe. If the lines are filled with liquid, then gas or vapour can enter them by:
(1) Changes of pressure drop and movement in the manometer allowing two-phase mixture to enter one or other of the pressure tappings.
(2) Flashing in the lines after a depressurisation.
(3) Pressure fluctuations in the tube causing a pumping action leading to gas ingress into the tappings. This phenomenon has been studied in some detail by Azzopardi et al (1977).

Apart from the obvious consequence of gas ingress in producing an unknown density of fluid in the tapping lines, the presence of gas bubbles in these lines can give a dramatic change in the frequency response with which pressure fluctuations can be measured. In vapour/liquid systems, the vapour entering the lines can be condensed by arranging a small cooling section just downstream of the pressure tapping point. However, the studies of Azzopardi et al showed that, in a typical system, the rate of condensation of the vapour bubbles entering the line was actually very slow.

If gas-filled lines are used, then liquid ingress can occur by:
(1) Changes in pressure drop and movement of the manometric fluid.
(2) Pressure fluctuations leading to liquid being pumped into the lines.
(3) Condensation of vapour in the lines.

Since the compressibility of the fluid in gas-filled lines is obviously much greater, the pumping action by pressure fluctuations is much worse than that for liquid-filled lines. Moreover, since in most two-phase flows, the liquid phase tends to wet the channel wall, capillary effects at the gas-liquid interface where the line enters the channel can be significant, particularly for small diameter lines. The consequences of liquid ingress into gas-filled lines are similar to those of gas ingress into liquid-filled lines. For liquid/vapour systems, an evaporator can be installed in the line just downstream of the tapping point, evaporating any liquid which enters. This is particularly useful for low latent heat fluids such as cryogenic fluids but, again, the rate of evaporation of the liquid is likely to be rather slow. On the whole, gas-filled lines are less satisfactory than liquid-filled lines although, of course, the offset at zero pressure drop (see equation (3.3)) is much less.

The performance of liquid-filled lines can be improved by using a balanced liquid purge system and a typical system is illustrated in Figure 3.2. Similar systems have been used by Simpson et al (1977) and by Thwaites et al (1976).

Depending on the pressure range to be covered, water-mercury manometers can be employed, or inverted water-manometers can be used (Hewitt et al (1962)). For greater sensitivity, water-carbon tetrachloride and water-kerosene manometers can be employed (McManus (1957), Serizawa (1974)). The problem with using any fluid-fluid manometer system is that an unexpected increase in pressure drop can lead to the transfer of the manometric fluid into the loop with, often, disastrous results. To overcome this difficulty, it is possible to introduce large diameter catchpots, in which the two fluids meet. For instance, Gouse and Hwang (1963), in conducting experiments with evaporating pentane, used pentane-water-air manometers in which the pentane was contacted with the water in large diameter reservoirs, such that the pentane-water interface moved only $\frac{1}{8}$ inch for a 12 foot change in water height.

An alternative to manometers for differential pressure measurement is the use of two wall-mounted pressure transducers, the signals from which are subtracted electronically. This system can give a fast response (thus

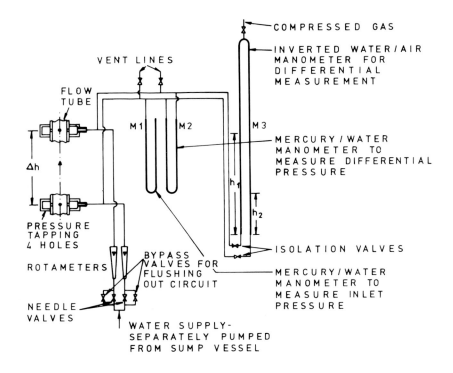

Figure 3.2: Liquid purge system for pressure drop measurement in two phase flow (Hewitt et al (1962)).

allowing the study of fluctuations in pressure drop), avoids the ambiguity in line content and has been used, for instance, by Chaudry et al (1965). The disadvantages of this system are that errors may be introduced in the subtraction of two large signals and also that the presence of gas bubbles near the transducer affects the frequency response.

The subject of pressure measurement in simulated reactor incidents is reviewed by Feinaer and Brockett (1963) and Ybarrondo (1975) and Brockett and Johnson (1976). A short guide to recently developed transducers is given by Votava (1974). The journal, Measurement and Data, has published a series of articles on pressure transducers giving advantages, disadvantages, ranges and manufacturer sources for various types. The types covered include potentiometric (Measurements and Data, Jan/Feb 1974), strain gauge (March/April 1974, March/April 1975), capacitive (May/June 1974), reluctive, inductive and eddy current (July/August 1973) and piezoelectric (Sept/Oct 1973).

Table 3.1 summaries the various forms of transducer which have been used; more details are given in the review by Brockett and Johnson (1976).

The use of transducers in measuring pressure fluctuations is discussed below. With absolute-pressure transducers, there is a problem of calibration range and it is sometimes more convenient to use differential-pressure transducers in a similar manner, referenced to a common source. Zuber et al (1967), in experiments on the evaporation of Freon, used six differential-pressure transducers along the channel, coupled to a common helium gas source, which was controlled to follow the system pressure.

The problem of subtracting large signals to give a differential pressure signal can be avoided by using a single, differential pressure, transducer with the two pressure tapping lines attached to two chambers separated by the diaphragm whose movement then indicates the differential pressure. Since only a small movement of the diaphragm occurs, DP cells have an advantage over manometers in that the movement of fluid into and out of the lines is minimised. The uses of various forms of differential pressure transducer and their relative merits, have been discussed by Lawford (1974); of the types of transducer itemised in Table 3.1, the strain-gauge and reluctance transducers are ideally suitable for differential pressure measurements but the capacitance and piezoelectric types are specifically unsuitable. The main problems in application of differential pressure transducers are as follows:

(1) Problems associated with tapping line content. These are similar to the problems found with manometric systems, and the reduction in frequency response as a result of gas content within the lines is particularly serious if one is attempting to measure pressure drop fluctuations. This problem is discussed, for instance, by Ybarrando (1975).
(2) The "offset" corresponding to zero pressure difference can sometimes present a serious problem. If one is attempting to measure pressure differentials which are small compared to the "offset" then the accuracy is necessarily limited since a DP transducer must be chosen with a range at least equal to the offset value. A solution to this problem, used by Webb (1970a) (1970b), is to use a compensating (eg CCl_4 - water) manometer as illustrated in Figure 3.3. Using this system, the "offset" is compensated by the head in the manometer.
(3) Problems associated with rig vibration. This can influence transducer response (see for example Weisman et al (1975)).

In summary, care has to be taken to minimise the effects of pressure-drop fluctuations, the effect of manometer lines and capillary effects. Pressure tapping lines should

TABLE 3.1

Classification of Transducer Type: With Advantages and Disadvantages

	Capacitance Type	Strain-Gauge Type	Reluctance Type	Piezo Electric Type	Potentio-metric Type	Magnetostrictive Type	Eddy Current Type
Sensitivity % full scale	0.01	0.3	0.1	0.001	0.25-1	0.5	0.01
Response time μs	20	10-100	200	2	2	2	10
Stability	Excellent	Good	Fair	Variable	Moderate	Good	Poor
Maximum Temp °F (standard types)	720	600	600	325	-	-	-
Cost	High	High	High	V.high	Moderate	V.high	High
Commercial availability	V.good	Good	Good	Good	Good	Poor	Moderate

"Potentiometric" — diaphragm moves slidewire
"Magnetostrictive" — magnetic properties varying with force

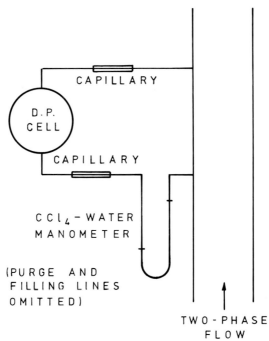

Figure 3.3: "Offset" compensation using a compensating manometer (Webb (1970a) (1970b)).

be filled with one of the phases, preferably the liquid phase. In general, the system for measuring pressure drop should be as simple as possible, whilst meeting the requirements. Where manometers are used, it should be possible to view the fluid within the lines if practicable so as to check for gas locks, etc. Alternatively, the compressibility of the fluid in the line can be checked by using acoustic methods. In the case of pressure transducers, or DP cells, periodic calibration is essential.

Development of pressure-difference measuring techniques appears to be sufficiently advanced to allow the selection of a suitable method for most applications. Possibly, more work is needed on the development of transducers for hostile environments and to withstand very rapid changes in pressure, so that they can be placed close to the pressure taps, thus shortening the lines.

Heat-Transfer Coefficient

Measurement of heat-transfer coefficient to a high degree of accuracy is difficult, even in single-phase flows. A vast literature has been built up in this area (see Bayley and Turner (1968)).

Two-phase heat-transfer can present more difficult

problems than single-phase heat-transfer, since the heat-transfer coefficient is often very high, necessitating a higher degree of accuracy in wall temperature measurement. Typical of two-phase heat-transfer investigations, for two-component (non-evaporating) flows, is that of Oliver and Hoon (1968) and of investigations of evaporating flows, that of Gouse and Dickson (1966).

If the fluid temperature is known, measurement of heat transfer coefficient involves the measurement of both the heat-flux and the surface temperature. Often, the local bulk temperature of the fluid is also unknown and that, too, has to be measured. For experiments on nuclear systems, the most common technique for measuring heat-flux is to heat the surface by means of joule heating, and to calculate the heat-flux from the current and voltage drop. However, in many experiments, heat is transferred through the surface from a heating or cooling fluid on the other side. In this case, the heat-flux is unknown and may vary in a complex way along the surface. An alternative to direct joule heating of the surface, is to use indirect electrical-heating and this is widely employed in experiments on emergency cooling. Here, although the time-average flux is known, the transient fluxes, resulting from changes in the cooling condition on the surface, are less easily specified.

Thus, with joule heating and with indirect electrical heating, the heat-flux in the steady state is known and the main problem is measurement of the wall temperature. With fluid-heated or-cooled systems, both the heat flux and the heat transfer coefficient have to be determined.

A general source book on temperature measurement is that of Hetzfeld, Dahl and Brickwedde (1962). A bibliography of temperature measurement for 1953-1969 is given by Freeze and Parker (1972). The most commonly used sensor for fluid temperature is the thermocouple and the main types of thermocouple employed are shown in Table 3.2.

TABLE 3.2

Main Types of Thermocouple used for Fluid Temperature Measurement

Main Types	Max. Temp. range ^{0}C	Sensitivity at 100^0C µV/K
Chromel/Alumel (Ni-Cr/Ni-Al)	-200 to 1400	41
Copper/Constantan (Ci/Ni-Cu)	-250 to 500	47
Iron/Constantan (Fe/Ni-Cu)	20 to 700	54
Platinum/Platinum 10% Rhodium	0 to 1500	7.4
Platinum/Platinum 13% Rhodium	0 to 1500	7.6
Tungsten/Tungsten 26% Rhenium	0 to 2300	3.3

A vast amount of literature is available on thermocouples. A useful source is the "Thermocouple Handbook" (CGS Thermodynamics (1974)) and further extensive information is given in the book by Early (1976). The effect of time and environment on the absolute accuracy and emf output of thermocouples is the subject of a paper by Keyser (1974). Figure 3.4 shows the various types of thermocouple junction used in fluid temperature measurement and gives also the approximate minimum response time of the various types.

A considerable source of error in thermocouple measurements is that resulting from temperature gradients in the leads. Care should be taken that conduction effects along the leads do not give ambiguity in the actual temperature of the junction; a good trick in measuring tube wall temperature is to wind the leads round the tube for one turn at the measurement level. Another common source of problems is where the leads pass through a region which is at a higher temperature than the junction. Illustrative of this problem are some results reported by Burnett and Burns (1977) and shown in Figure 3.5. Thermocouples with their junction in boiling water had 7 metres of their leads placed in a muffle furnace at 900°C. The indicated temperature varied with time and from thermocouple to thermocouple as shown. This effect probably arises from inhomogeneities in the leads. The uncertainties of using thermocouples in these circumstances can be overcome to some extent by designing multiple sensors combining two thermocouples and a platinum resistance sensor as illustrated in Figure 3.6. The platinum resistance sensor calibrates the response of the thermocouples and the whole device can be mounted in a conventional sheath (Corradi (1977)).

Other devices used for temperature measurement include:
(1) *Resistance bulbs*. These are typically of platinum, can be used over the range 20-1400K, and have a sensitivity of 0.2-20 ohms/K. The response time is normally rather slow (1-10s). The use of resistive elements is reviewed by Kunz (1968).
(2) *Thermistors*. These are semi-conductor devices operating in the range 50-480K giving typically a 4% change in resistance per degree K. With a small enough bead, a

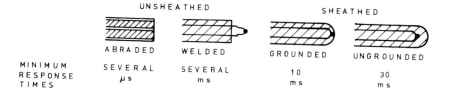

Figure 3.4: Types of thermocouple used in fluid temperature measurement.

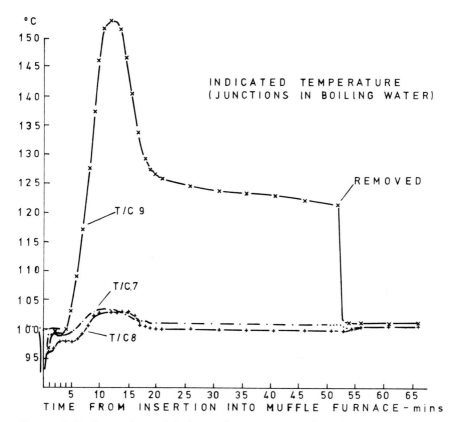

Figure 3.5: Variation of indicated temperature for thermocouples with 7m of lead inserted in a muffle furnace at 900°C (Burnett and Burns (1977)).

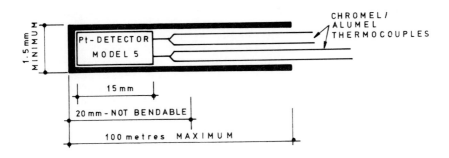

Figure 3.6: Triple temperature sensor designed to eliminate lead effects (Corradi (1977)).

response time of 3ms is possible. A review of the use of resistance and thermistor thermometry is given by the journal Measurement and Data (March/April 1974 and March/April 1975).
(3) *Fluid filled thermometers.* These can be liquid filled (operating in the range 230-670K) or gas filled (4-1050K). Fluid filled thermometers have a very slow response time.
(4) *Bi-metallic thermometers.* These can operate in range 90-1000K but, again, have a rather slow response.
(5) *Optical pyrometers.* These are particularly suitable for very high temperatures but problems of obtaining a suitable view factor often inhibit their use. A discussion of the use of optical pyrometry in heat transfer is given by Kunz (1977).

Further comparitive information on the various sensors is given by Brockett and Johnson (1976).

Considering first the problem of wall-temperature measurement, we see that this is one of the most severe difficulties in measuring the heat-transfer coefficient. The problem is that the positioning of a thermocouple in the surface automatically distorts the flux profile for both electrical- and fluid-heating. In the case of fluid-heating or cooling, the thermocouple will generally pass through the fluid and cause a local disturbance in the boundary layer, which again affects the heat-flux. Both these considerations have caused a gradual movement towards smaller and smaller thermocouples.

A survey of wall-temperature measurement problems is given by Watson (1964). Walker and Rapier (1965) examine, theoretically, the influence of the insertion of a thermocouple in a tube wall, for both internal heat generation and for heat transmitted from a heating fluid. For electrical heating, the thermocouple can be set on the wall of the channel and the temperature on the other wall estimated from the heat-flux and thermal conductivity of the tube material. Often, the temperature drop through the wall is of the same order, or greater than, the temperature drop in the heat-transfer from the wall to the fluid. Thus, small errors in wall thickness or thermal-conductivity measurement can lead to large errors in heat-transfer coefficient. There is an obvious advantage in using thin, high thermal-conductivity walls in such measurements. Bertoletti et al (1964) measured the temperature difference directly, by having thermocouples attached to an electrically heated zone and to an immediately adjacent unheated zone, the latter thermocouple ostensibly giving the fluid temperature.

In measuring the outer surface-temperature of an electrically-heated tube, it is often difficult to avoid pick-up problems, particularly with AC heating (Buchberg et al (1951)). Sheathed thermocouples can be used (see for instance Scruton and Chojnowski (1978)). At Harwell, the preferred method is to isolate the thermocouple from the wall by using very thin anodised aluminium foils, layers of insulation and guard heating are then placed around the tube, in order to minimise the temperature gradient through the foil and to obtain accurate values of wall temperature. Similarly, Laverty and Rohsenow (1967) used mica insulation. For rapid response, direct attachment of the thermocouple wires is necessary and a three-wire system can be used. Another alternative is to use the tube wall itself as a resistance thermometer. For normal materials, this gives rather limited accuracy but it is worth noting that high accuracy can be obtained by using special techniques. For instance, Green and Hauptmann (1970) covered a cylinder with gold-coated platinum films to give a reproducible high temperature-coefficient of resistivity and, thus, an accurate determination of surface temperature. Thin-film wall-temperature sensors are also described by Bickel and Keltner (1973).

For fluid-heated or cooled systems, it is necessary to set the thermocouple in the wall of the channel and much development has been described in this area. For instance, Elliot and Dukler (1970) have developed a technique, as illustrated in Figure 3.7, for locating the thermocouple junction precisely at the surface. A chromel plug, with an insulated constantan wire passing through it, is inserted into the surface and a junction is made by polishing the end of the plug, thus creating a junction precisely at the tube wall. Such a device does not avoid the problem of heat-flux distortion, mentioned above, although this is minimised by introducing the plug at an angle as shown in Figure 3.7. In their studies of the circumferential variation of heat transfer coefficient around the outside of a tube in condensation, Nicol et al (1978) used a similar principle. Here the thermocouples were set in small slots in the tube which were then sealed with a fillet of brass soldered in position. A commercial design (NANMAC), of the abraded thermocouple type, incorporates a plug body made of the same material as the wall, to minimise the problem of heat-flow distortion discussed here.

A number of circumstances exist in which very precise measurements of surface temperature are required and one of these is in the measurement of condensation coefficient. This problem is discussed by Wilcox and Rohsenow (1969) and they suggest that the non-unity values of condensation coefficient, which had been previously measured, were, in fact, due to errors in the measurement of wall temperature. In their own experiments, they estimated water temperature by having a series of thermocouples in a very thick wall, placed normal to the direction of heat-flow. A plot of temperature versus distance was then extrapolated to the surface to give the surface temperature. A somewhat similar method was used by Necmi and Rose (1977) and Grigoriev et

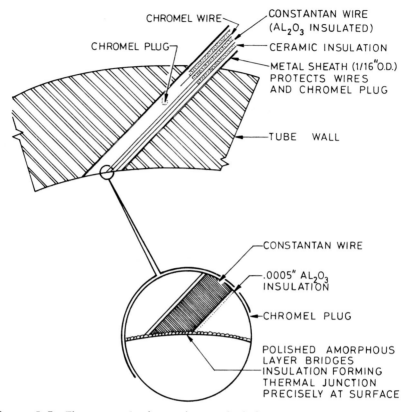

Figure 3.7: Thermocouple insertion method for wall-temperature measurement (Elliott and Dukler (1970)).

al (1978). Wilcox and Rohsenow make the interesting observation that values of condensation coefficient greater than unity would be rejected as being physically unreasonable. Thus, only values less than unity would be published and this would give a distorted picture!

In the case of fluid-heated systems, the heat-flux has to be determined in addition to the wall temperature. This flux can often be obtained by means of a heat-balance on the respective fluid streams. This is not particularly satisfactory in determining local heat-fluxes in cases where the heat-transfer coefficients change in an unknown way (ie those cases where the coefficient needs to be measured!). Thus, measurements of local heat-flux are required. Such fluxes can be measured by measuring the temperature gradient in the channel wall. However, unless extremely thick walls are used (eg Wilcox and Rohsenow (1969)), errors can occur due to the distortion of the flux-pattern. Other measurements, using the temperature-difference method for determination of heat flux, are reported by Pletcher and McManus (1969) and Butterworth et al (1974). Measurements were made of local heat-flux in condensation in a horizontal tube. Two thermocouples were placed in the wall as illustrated in Figure 3.8. One thermocouple was set into a groove in the outer surface and the other was placed in a hole drilled at a small angle to the outside of the tube. The difference between the two temperatures was used to determine local heat-flux. It was necessary to calibrate the system using single-phase transfer on both sides of the tube. This also allowed the temperature reading to be corrected to allow determination of wall temperature.

The problems of wall-temperature measurement have led some authors to use indirect methods of determination. For instance, Edwards et al (1974) report the use of a phosphorescence technique. The intensity of phosphorescence from a local spot on the surface was a function of temperature. Another example of indirect temperature measurement is the work of Newby (1971), in which an infra-red radiometer was used to measure the surface temperature of a liquid film. An accuracy of about ±0.5°C was claimed.

In heat transfer to solid-gas systems, for example, fluidized beds, heat transfer between an inserted element

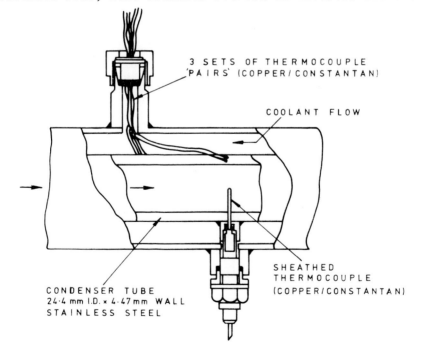

Figure 3.8: Measurement of wall heat-flux and surface temperature in horizontal tube condensation (Butterworth and Pulling (1974)).

and the gas-solid mixture is often determined by using elements such as spheres, with heat generation near the centre and temperature-gradient determination to evaluate heat-flux. Examples of this kind of work are that of Klimenko et al (1970a) (1970b), Nodak (1970), Babukha and Sergeyev (1970) and Botterill and Sealey (1970).

Measurement of received heat flux is of considerable importance in radiative heat-transfer and localised heat-flux meters have been widely developed in this field. Examples of such developments are those of Nerem and Stickford (1964), Bakum et al (1968), Schulte and Kohl (1970), Rudenko (1971) and Kirsanov et al (1973). Since, with flux meters of this type, the surface temperature is also often measured, they would seem to be ideal for measuring heat-transfer coefficients. However, there is a serious drawback; the heat-transfer coefficient depends on the development of the thermal boundary-layer. Thus, with very short heated lengths, the heat-transfer coefficient is extremely high, because the thermal boundary-layer has not developed. Therefore, the heat-transfer coefficient measured by a local heat-flux meter would not be representative of the system, unless the local heat-flux was exactly identical, in the heat-flux meter, to that applying in the surrounding surface.

In an attempt to develop a heat-flux meter which did not present the difficulties described above, Gardon (1960) evolved the heat-flux meter illustrated in Figure 3.9. A small copper tube (B) is inserted through the surface (P) and the end of it closed off with a thin constantan foil (F) at the centre of which is connected a copper wire (W). The temperature difference between the outside and the

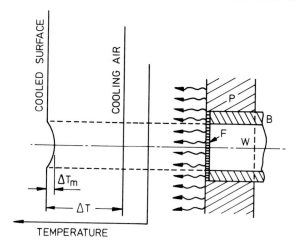

Figure 3.9: Principle of operation of heat-flux meter (Gardon (1960)).

centre of the foil (ΔT_m) is determined from the emf between B and W and is a direct measure of the local heat flux. By using an insert of very small dimensions, accurate measurements can be made of the heat flux locally. Further developments of this kind of instrument are reported by Northover and Hitchcock (1964), Hager (1965) and Landram (1974).

An alternative method for measuring heat-transfer coefficient, is to measure the transient response of the temperature of a body, or part of a body, cooled by a fluid stream. A transient method for local heat-flux and heat-transfer coefficient measurements has been developed by Oka et al (1970). These authors used a metal insert in an insulated fin. The metal insert was preheated and its temperature decay determined to give a measure of the local heat transfer coefficient. Here, the heat flux pattern must be undisturbed over the surface and the results must be interpreted with care.

A specific difficulty is encountered in the case of the use of electrical heating for film boiling and transition boiling heat transfer measurements. Here, the temperature excursion on entering the film boiling region is often sufficient to damage the test section. A way round this difficulty has been devised by Ralph et al (1977) who used a test section as illustrated in Figure 3.10. A series of copper blocks are pre-heated to the required temperature and then water is introduced at the bottom of the test section. Film boiling occurs and measurement of the temperature distribution in the blocks as a function of time allows the determination of the local wall temperature and heat fluxes (and hence heat transfer coefficients). Rewetting of the surface begins at the bottom of the test section and a feature of systems shown in Figure 3.10 is that the lower block can be pre-heated to a higher temperature than the middle blocks thus slowing down rewetting and allowing coefficients to be determined on the centre blocks for much lower wall temperatures. Other manifestations of the same principle are reported by Ralph et al (1978) and Ganic and Rohsenow (1977).

Another technique for measurement of heat-flux in fluid-heated or cooled systems has been developed by Michiyoshi et al (1969). These authors positioned a thermocouple close to the wall in a single-phase fluid (eg the coolant stream in a condenser experiment) and measured the magnitude of the local fluctuations in temperature, which varied linearly with the heat-flux and which could be used to give a local heat-flux at that point. Another technique using temperature fluctuation is described by Mumme and Lawther (1973). When the heat-transfer surface is thin, the response of the surface to fluid temperature fluctuations is dependent only on the heat-transfer coefficient and the thermal capacities of fluid and solid. Measurement of the attenuation of either cyclic or random fluid-

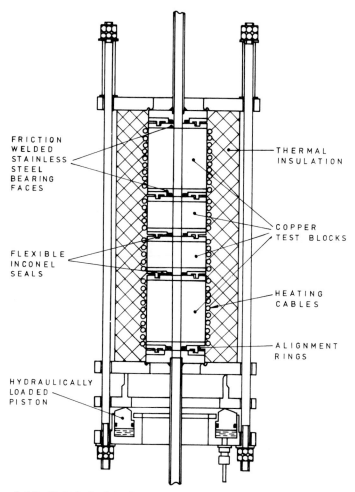

Figure 3.10: Multiple-block test section used in the film boiling studies of Ralph et al (1977).

temperature variations can then give the heat-transfer coefficient.

Finally, one might mention the use of photographic techniques to determine heat-flux in multiphase systems. Thus, the evaporation of a droplet can be observed and its rate of evaporation determined. An example of work of this type is that of Adams and Pinder (1972).

Measurement of two-phase heat-transfer coefficient can present more difficult problems than the measurement of single-phase coefficients since the heat-transfer coefficient is often very high, necessitating a higher degree of accuracy in wall-temperature measurement. With joule heating, the heat flux in the steady state is known and the main problem is measurement of wall temperature. With fluid-heated or fluid-cooled systems, both the heat-flux and the heat-transfer coefficient have to be determined. There is no universally-applicable method for measuring heat-trasnfer coefficients in multi-phase flow systems. Each system must be considered on its own merits and the appropriate techniques selected. More development work is reqaired on the measurement of local heat-transfer coefficients, particularly for fluid-heated systems.

Mass-Transfer Coefficient

Inter-phase mass transfer is a massive subject in its own right and it is beyond the scope of the present review to cover it in any detail. Mass transfer coefficients may be important in two-phase flow design in the following areas:
(1) In designing mass transfer systems, such as pipe-line contactors. The general subject of such contactors has been surveyed by Alves (1970).
(2) Mass transfer may be an important limitation on heat-transfer in the condensation of multi-component vapours, and of vapours with incondensible gases present.

It is also of interest to mention, in this section, two modern techniques for measurement of mass transfer which, though not of direct design interest, may find an increasing application in studying two-phase flow phenomena. These techniques are the thin-film technique and the electro-chemical technique, respectively.

Inter-phase mass transfer is strongly influenced by flow pattern as illustrated, for instance, by the work of Mah and Golding (1971). A wide range of data exists for mass transfer in bubble columns and this is exemplified by the work of Shulman and Molstad (1950), Yau et al (1968), Deckwer et al (1974), Kress (1972) and Kawagoe et al (1975). Absorption of carbon dioxide in horizontal slug-flow was studied by Gregory and Scott (1968) and liquid-side coefficient measurements in vertical upwards flow are reported by Tomida et al (1976) and Shilimkan and Stepanek (1977). In the latter tests, carbon dioxide was absorbed into a buffer solution of sodium carbonate and sodium bicarbonate. The absorbed carbon dioxide reacted with the sodium carbonate in such a way that the concentration of carbon dioxide in the bulk liquid was essentially zero, though the reaction rate was such that no appreciable amount of the absorbed component reacted in the diffusion film at the surface of the liquid. In this case, diffusional resistance in the liquid phase is controlling and the mass transfer obeys the formula:

$$Ra = k_L \, a \, C^* \qquad (3.6)$$

where R is the transfer rate per unit intefacial area, a is the interfacial area per unit volume, k_L the mass transfer coefficient and C^* the equilibrium CO_2 concentration at the interface. Shilimkan and Stepanek found that the volumetric mass transfer coefficient ($k_L a$) increased with superficial liquid velocity and passed through a maximum with increasing gas velocity as shown in Figure 3.11. The maximum could reflect a regime transition to annular flow where there could be a reduction of interfacial area.

In all the above studies, the mass-transfer rate was determined by measuring the concentration of the dissolving gas in the liquid phase. An alternative technique is described by Calderbank and Johnson (1970) who studied the mass transfer from a single bubble rising in a tube. The tube was sealed and mass transfer from the bubble did not change its volume, but dissolution of the gas caused a change of pressure which could be measured and the rate of mass transfer determined. A further discussion of methods is given by Garbarini and Tien (1969). In liquid-film flows, the presence of surface waves has a large effect on the gas-to-liquid mass transfer (Konobeev et al (1957)) and the effect of interfacial waves is particularly marked inthe case of mass transfer in annular flow, where a further complication is the existence of dispersed, entrained droplets within the flow. Schneiter (1966)

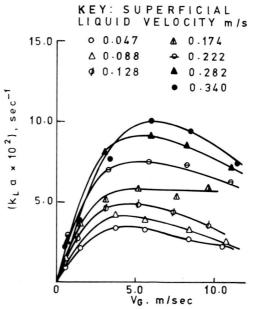

Figure 3.11: Volumetric mass transfer coefficients obtained by carbon dioxide absorption. Data of Shilimkan and Stepanek (1977).

determined mass-transfer coefficients in air-water flow by measuring the change in specific humidity of the air flowing with the water. Anderson et al (1964) have studied mass transfer in the gas core of horizontal annular flow by injecting ammonia gas and studying its absorption by the liquid film. Allowance was made for the effect, on mass transfer, of droplet interchange between the film and the gas flow. The basic assumption of this technique was that the mass transfer was gas-phase controlled. Coefficients up to an order of magnitude greater than those for smooth tubes were observed, presumably due to the influence of interfacial waves. However, considerable problems of definition of mass-transfer coefficient may exist in this type of experiment, due to circumferential variations of the concentration of the dissolving gas in the liquid film.

Studies of simultaneous heat and mass transfer to freely falling and to fixed droplets are described by Kulic et al (1975) and Kulic and Rhodes (1977) respectively. Water droplets were contacted with steam/air mixtures; in the experiments with fixed droplets. The droplet diameter and temperature were measured as a function of time and Figure 3.12 shows a typical set of data. The heating process was mainly controlled by the mass transfer of water vapour to the interface, though, with no internal mixing within the droplet, the internal resistance could have been significant. Studies of the vaporisation and combustion of single fuel droplets are reported by Ohta et al (1975); here, the droplet motion and evaporation was studied photographically in artificially produced and naturally occuring turbulence

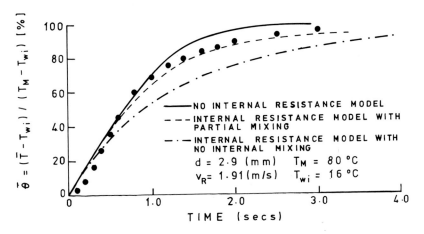

Figure 3.12: Variation of the mean temperature $\bar{T}$ of a water droplet at initial temperature T_{wi} as a function of time of exposure to steam/air mixture at temperature T_M. Results of Kulic and Rhodes (1977).

fields. An ingenious method for studying mass transfer to suspended droplets is described by Mensing and Schugerl (1970). Liquid scintillators are added to the liquid from which the droplets are formed and tritium-labelled tracers are added to the gas phase. When the tracer is transferred to the droplet, scintillations occur in the droplet itself and are detected.

Mass transfer is of considerable technical importance in fluidised bed systems and there have been many measurements; these are typified by the result reported by Morooka et al (1977), Hsiung and Thodos (1977) and Koloini et al (1977). Hsiung and Thodos studied gas fluidised beds in which the particle-to-gas mass transfer coefficients were studied by mixing napthalene particles with the bed and measuring the concentration of napthalene in the outlet gas. Koloini et al measured the rate of ionic uptake by the bed particles to measure liquid-to-particle mass transfer coefficients in liquid fluidised beds and Morooka et al's measurements (using transient tracer techniques) were concerned with "bubble phase" to "emulsion phase" interchange in gas fluidised beds.

Mass-transfer analogy techniques have been widely developed for evaluating transport processes, particularly from complex surfaces. One of the most widely used is the thin-film technique, in which a coating of material is applied to the surface and the rate of removal of the surface coating, by the surrounding flowing fluid, is determined. For instance, a thin coating of naphthalene can be applied to the surface and the overall coefficient determined by measuring the outlet concentration of naphthalene. Local coefficients can be determined by measuring the local changes of thickness and development work has been carried out on the methods for depositing a uniform thickness of film on the surface (Neal et al (1970), Neal (1975)). Recently, Kapur and MacLeod (1974) (1975), have developed the use of holography in determining the changes in surface profile induced by the evaporation of the thin deposited film. A hologram is made of the initial surface and of the surface after mass transfer has taken place. Interference between the two holograms is a direct indication of the changes in surface profile, from which the mass-transfer coefficient can be deduced with a high degree of accuracy. A typical fringe-pattern, obtained by this technique, is illustrated in Figure 3.13, for the case of mass transfer of ethyl salicylate from a silicon-rubber coating on a flat plate exposed to a normal water-jet. It is possible that developments of this technique could be applied in two-phase flows, for example, in studying local phenomena around grids.

Another technique for the measurement of wall-to-fluid mass transfer (and, by analogy, heat transfer), which has achieved prominence in recent years, is the so-called "electro-chemical method". In this method, the flowing

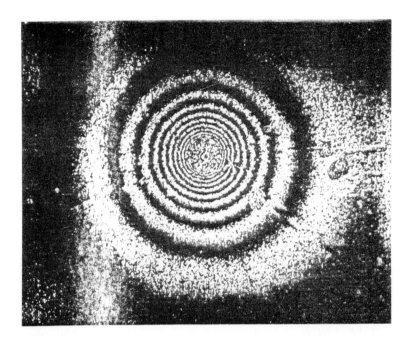

Figure 3.13: Interference fringes obtained by holographic interferometry, recording the mass transfer of ethyl salicylate from a silicon-rubber coating on a flat plate exposed to a normal water-jet (Kapur and MacLeod (1975).

fluid contains an electrolyte and the current passing from the surface into the fluid is limited by diffusion of ions from the bulk fluid, through the boundary layer, to the surface. A detailed review of the electro-chemical method is given by Mitzushina (1971).

The section of surface on which the mass transfer coefficient is to be measured forms one electrode and the other electrode, placed elsewhere in the system, is made much larger and does not control the current. It is convenient to use a redox reaction, for example the ferricyanide/ferrocyanide reaction. Here the electrodes are usually made of nickel, with the cathode being the electrode normally chosen as the mass transfer surface. The electrode reactions in this case are at the cathode:

$$Fe\ (CN_6)^{3-} + \varepsilon \rightarrow Fe\ (CN_6)^{4-} \qquad (3.7)$$

and at the anode:

$$Fe\ (CN_6)^{4-} \rightarrow Fe\ (CN_6)^{3-} + \varepsilon \qquad (3.8)$$

As the voltage is increased, the concentration C_W of ferricyanide ions at the cathode surface falls and ultimately reaches zero. From this point onwards, the current is governed by diffusion of ferricyanide ions through the boundary layer and is constant at a value I_{lim} over a wide range of voltage as illustrated in Figure 3.14.

One may define a mass transfer coefficient $\bar{k}_c$ for the transfer of ions across the boundary layer and this is related to the limiting current by the expression:

$$\frac{I_{lim}}{AF} = \bar{k}_c (C_B - C_W) = \bar{k}_c C_B \qquad (3.9)$$

where C_B is the concentration of ferricyanide ions in the bulk fluid and F is the Faraday number. In operating with systems containing air, care has to be exercised to avoid oxidation of the solution. Such oxidation can be minimised by pre-saturation of the solution with nitrogen.

A discussion of the application of the technique to two-phase flows is given by Sutey and Knudsen (1969). The technique is particularly powerful in looking at complex geometries, as exemplified by the work of Dawson and Trass (1972) on rough surfaces. It has also been found useful in investigating the surface-to-liquid transport processes which occur in bubble growth during boiling (Bode (1972)) and in the determination of fluid-to-wall mass transfer in liquid fluidised beds (Smith and King (1975)).

The high Schmidt numbers associated with the diffusion in the ferrocyanide solutions used in this technique, mean that the mass-transfer coefficients observed, are determined by processes occurring immediately adjacent to the wall and depend on the velocity profile in this region which, in turn, is governed by the wall shear-stress. The

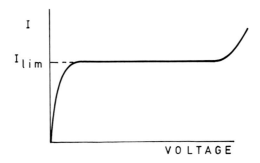

Figure 3.14: Current/voltage curve for diffusion controlled electrolysis method.

technique has been extensively used as a means of determining wall shear-stress and will be referred to again, in this context, in Chapters 6 and 7.

A large variety of technique exists for measuring two-phase mass transfer. Mass-transfer analogy techniques, including thin-film deposition and the electro-chemical method, may find increasing applications in two-phase flow.

Further work should be encouraged on the application of the electro-chemical method in two-phase flow. This could throw considerable light on the various phenomena occurring, particularly those associated with the components of pressure drop.

Mean Phase Content (Void Fraction)

The application of void fraction measurements were described in the introduction to the Chapter. The importance of these measurements has led to extensive development work. The main methods used are radioactive absorption and scattering, impedance and volume measurement. A detailed discussion of these principal techniques is given first below and then a brief outline is given of alternative techniques.

Review of void fraction measurement techniques are given by Williamson (1970) (with particular reference to cryogenic systems) and by Lahey (1978).

Radioactive Absorption and Scattering Methods

Probably the most widely used technique, for the measurement of void fraction, is measurement of the attenuation of a beam of gamma rays in the flow. Attenuation of gamma beams in passing through materials occurs by three distinct processes, the relative importance of which varies with the gamma photon energy and with the attenuating material. The attenuation processes are:

(1) *Photoelectric effect:* Here the gamma-ray photon gives all its energy to an atom, causing the ejection of an electron from an inner orbit.

(2) *Pair production:* The photon creates a positron-electron pair and is absorbed in the process. The positron is subsequently annihilated with the production of two 0.51 MeV photons; since the pair production is of importance only at high gamma energy, the secondary photons produced are absorbed much more readily than the incident beam and, effectively, complete absorption can be assumed.

(3) *Compton effect:* Here, the gamma photon interacts with an atomic electron, gives some energy to it and proceeds with lower energy and altered course. The energy E' of the scattered photon (MeV) is related to the initial energy E and the scattering angle θ by the relationship:

$$E = \frac{E}{1 + 1.96E\ (1 - \cos\theta)} \qquad (3.10)$$

The absorption of a collimated beam of initial intensity I_0 (photons/m²s) is described by an exponential absorption beam as follows:

$$I = I_0 \exp(-\mu z) \qquad (3.11)$$

where μ is the linear absorption coefficient and z the distance travelled through the absorbing medium. Data on absorption coefficients are available in standard nuclear handbooks (eg Blatz (1959) and Etherington (1958)) and are often presented in terms of μ/ρ where ρ is the density of the absorbing material. Typical data for μ/ρ are shown in Figure 3.15, for lead.

The dominant role of the photoelectric effect for attenuation in lead at low energy will be noted. However, the relative contributions of the respective mechanisms varies with the substance. For instance, with attenuation in iron, the Compton effect is dominant over a wide range of energies. Since the Compton effect produces secondary photons which may themselves be detected, differences exist between broad beam and narrow beam systems. The original incident photons can be distinguished from the secondary photons by using gamma spectrometry, but this

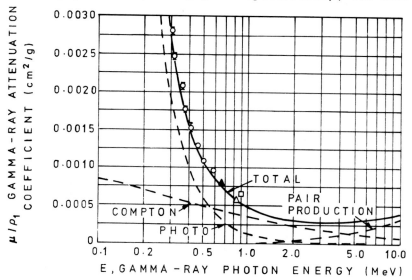

Figure 3.15: Gamma-ray attenuation coefficient for lead (from Blatz (1959).

can be expensive. The best solution is probably the use of in-situ calibration with measurements of the intensity with the tube full of the gas or vapour (I_G) and full of the liquid phase (I_L). Assuming an exponential absorption, the vapour fraction (α) is then derived from the received intensity I from the equation:

$$\alpha = \frac{\ln I - \ln I_L}{\ln I_G - \ln I_L} \qquad (3.12)$$

However, the complex nature of the attenuation process should always be borne in mind in designing gamma absorption systems.

Tables 3.3 lists the main types of sources which have been used in void fraction measurements. The most common are Thulium-170 and Cesium-137 and these isotopes have the advantage of giving a largely monenergetic emission. The absorption coefficients for water and iron are given for guidance; however, for the reasons given above, each installation should be calibrated in situ as previously described. The "half distance" shown in the table is the distance travelled to reduce the radiation intensity by a factor of two.

A brief introduction to the use of radiation attenuation techniques in two phase void measurements is given by Schrock (1969). This method has been widely used and developed and studies using it on two-phase flow in tubes, include those of Bailey et al (1955), Miropolsky (1955), Marchaterre (1956), Isbin et al (1957), Gibson et al (1957), Dixon (1957), Ziegler and Chiangkasri (1958), Petrick and Swanson (1958), Hooker and Popper (1958), Richardson (1958), Miropolsky and Shneyerova (1960), Isbin et al (1959), Vinokur and Dilman (1959), Bartolomei et al (1960), Fohrman (1960), Foglia et al (1961), Haywood et al (1961), Condon and Sher (1961), Hoglund et al (1961), Sher et al (1962), Heineman et al (1963), Jansen and Kervinen (1964), Fauske (1965), Ferrell and Bylund (1966), Milliott et al (1967), Bartolemei and Chanturiya (1967), St Pierre and Bankoff (1967), Lottes (1967), Henry (1968), Gardner et al (1970), Piper (1971), Hancox et al (1972), Ochial et al (1972), LeVert and Helminski (1973), Abuaf et al (1978), Jones (1977), Stephens (1977), Thomas et al (1977) and Lockett and Kirkpatrick (1975). Measurements in annuli are reported by Evangelisti and Lupoli (1969) and Zakharova et al (1970) and measurements in tube bundles are reported, for instance, by Gustafsson and Kjellen (1971). The use of gamma techniques in converging-diverging nozzles is reported by Vogrin (1963), Hammitt et al (1964) and Smith et al (1964). Baumeister et al (1973) report the use of the technique in studying air-water critical-discharge phenomena. A typical commercially available instrument for gamma densitometry is described

TABLE 3.3

Gamma-Ray Sources used in Void Fraction Measurements

Source	Typical reference to use	Gamma energy keV	Half life	Mass absorption coefficients $(\mu/\rho)\,cm^2/g$		Linear absorption coefficients $(\mu)\,cm^{-1}$		"Half distance" cm	
				Water	Iron	Water	Iron	Water	Iron
Cesium-137	Lassahn (1975)	662	30 years	0.086	0.073	0.086	0.57	8.1	1.2
Thulium-170	Petrick and Swanson (1958)	84	127 days	0.18	0.5	0.18	4	3.8	0.2
Irridium-192	Bailey et al (1955)	317, 468, 605	74 days	–	–	–	–	–	–
Selenium-75	Isbin et al (1957)	265, 136, 280 24-58	120 days	–	–	0.12*	–	5.7	–

* Experimentally measured by Isbin et al

by Measurements Incorporated (1975).

An alternative to using gamma rays is to employ X-rays and, in recent years, this method is becoming more popular with the development of improved X-ray sources. Reports on the use of X-rays for void fraction measurement include those of Anderson and Eno (1960), Kemp et al (1959), Martin and Grohse (1960), (1961), Kendron et al (1963), Pike et al (1965), Miropolsky and Shneyerova (1962), Schrock et al (1966), Zuber et al (1967), Walmet and Staub (1969), Martin (1972), Lahey (1977) and Smith (1975).

A detailed discussion of the errors arising in radiation-attenuation void measurements is given by Piper (1974). The main difficulties and problems arising are as follows:
(1) There are the ordinary difficulties of handling radiation. These problems are discussed in general terms by Clark (1963) and Evans (1963), and both these sources are useful reading before attempting to use the techniques.
(2) There is a fundamental inaccuracy in the measurement of void fraction due to the normal photon statistical fluctuations. This error can be minimised by using long counting times or strong sources. The statistical error is inversely proportional to the square root of the number of counts. A detailed discussion of statistical errors is given by Piper (1974). Piper derived an expression for the required strength of a source to produce a given accuracy. Suppose that a count rate of R_0 is obtained for a gas or vapour filled channel and R_1 for a liquid filled channel. The range R_0 to R_1 can be regarded as "fullscale" for the detector and errors can be expressed as "percent fullscale" (% FS). The standard deviation σ in the count rate is given by:

$$\sigma = \sqrt{R/\tau} \qquad (3.13)$$

where τ is the counting time and the %FS error corresponding to K standard deviations is given by:

$$\%FS \text{ error} = \frac{100K \sqrt{R/\tau}}{R_0 - R_1} \qquad (3.13)$$

Piper calculates the source strength in curies required to obtain a given %FS error as follows:

$$\text{Curie Strength} = \frac{4h^2 \exp(\mu_M z_M)}{3.7 \eta \gamma \tau r^2 \times 10^{10}} \left[\frac{100K}{(1-\exp(-\mu z_{max}))\%FS} \right]^2$$

$$(3.14)$$

where h is the distance from the source to the detector, r the collimation radius, η the detector efficiency, μ_M and z_M the absorption coefficient and the distance travelled through the metal walls, γ the fraction of the source yield emitted as gamma photons, μ the absorption coefficient of the liquid and z_{max} the maximum distance the beam travels through the liquid. As will be seen by applying equation 3.14, strong sources and/or long counting times are required to obtain high precision. The problem is most severe in the cases of gamma-absorption measurements where shielding considerations limit the source strength. Martin (1972) plots the percentage error in void fraction, for his X-ray absorption method, and the plot is reproduced in Figure 3.16. It will be seen that the largest errors are likely to occur in local void-fraction measurements, particularly at low void-fractions. To overcome these difficulties with γ absorption, source strength of up to 30 curies have already been employed.
(3) Influence of void orientation. Petrick and Swanson (1958) have shown that there are two limiting cases. When the liquid and vapour exist in layers parallel to the beam, the void fraction is determined by:

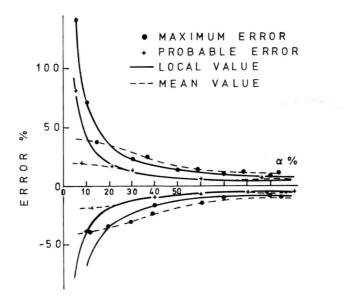

Figure 3.16: Errors in determination of void-fraction for mean and local void-fraction measurements (Martin (1972)).

Mean Phase Content (Void Fraction)

$$\alpha = \frac{I_x - I_L}{I_G - I_L} \qquad (3.15)$$

where I_x is the measured intensity and I_L and I_G are the intensities measured for the case of the tube full of liquid and full of gas, respectively. For the opposite case, when the liquid and vapour exist in layers perpendicular to the beam, α is given by:

$$\alpha = \frac{\ln(I_x/I_L)}{\ln(I_G/I_L)} \qquad (3.16)$$

Wall-flows fall between these two extremes but it is usually found that the second equation more closely represents the actual situation. Experiments with plastic models are reported by Petrick and Swanson (1959), Richardson (1958), by Kondic and Hahn (1970) and by Dementev and Skachek (1973) and these show that considerable errors occur in the application of the second equation, in situations of separated flow (annular flow, sub-cooled boiling). More accurate data can be obtained by determining the void-fraction distribution and taking an average (see below). The errors are minimised when the total absorption of gamma rays even with the channel full of water is small. In this case, the absorption approximates to linear and the orientation effects disappear. This can be achieved by using a hard gamma but the counting statistics are less favourable.

(4) Effect of tube wall on averaging. In making void-fraction measurements in a tube, the beam passes through variable metal-wall thickness. Using a single beam to take an average over the whole tube, there is a stronger weighting for those parts of the tubes where the beam-path lengths through the metal wall are lower. This problem can be overcome, to some extent, by the use of specially designed, shaped collimators which are placed between the source and the tube (Ferrell and McGee (1965), Lottes (1967)).

(5) Time fluctuation effects. Two-phase flows are often time-varying and an extreme example of this would be slug flow, where the void fraction fluctuates periodically from nearly zero to nearly unity. If the absorption is exponential, then, clearly, the average signal does not represent the mean void-fraction. The problem of inaccuracies arising from time fluctuations is discused by Harms and Forrest (1971) and Harms and Laratta (1973). A possible way of reducing these inaccuracies is suggested by Le Vert and Helminski (1973). They propose that two gamma-ray beams of

widely different energy should be used. The ratio of the time-averaged intensities from the two beams gives a measure of void fraction and the use of the ratio gives a result for mean void fractions in a time-varying flow, which is closer to the true value than the equivalent mean obtained from a single beam. Barrett (1974) discusses the relative effects of counting-statistics errors and time-fluctuation errors. He describes a method of designing experiments to minimise overall errors. An analytical study of time-fluctuation errors is given by Jeandrey and Pinet (1978). Alternatively, a strong source can be used whose output can be monitored continuously, linearised and then averaged. Also, if a hard source is used, then the absorption, even with the channel full of water, is small and the response approximately linear. In this latter case, an optimum has to be struck minimising the sum of errors due to counting statistics and errors due to void fluctuations and void orientation.

(6) Non-monoenergetic nature of radiation. As shown in Table 3, many sources have a range of gamma energies and also as described above, photons of reduced energy are generated by the Compton effect and can be picked up by the detector. The combination of these effects can often lead to abiguity in the interpretation of the signal and should be carefully evaluated for the particular system if high accuracy is desired.

The use of β ray absorption for the measurement of void fraction is also feasible, though there are severe limitations, due to the absorption of the β rays in the channel wall. However, the technique can be particularly useful where only a small amount is present (eg a dispersion phase in steam). Work on β ray absorption is reported by English et al (1955), Costello (1959), Styrikovich and Nevstrueva (1960), Korsunskii et al (1960), Cravarolo et al (1961), Perkins et al (1961) and Zirnig (1978).

Neutron absorption methods for measuring void fraction have been investigated by Dennis (1957), Ager-Hanssen and Døderlein (1958), Sha and Bonilla (1965), Harms et al (1971) and Banerjee et al (1978a). Dennis used a radium-beryllium source for out-of-pile evaluation of void fraction and this turned out to be less satisfactory than gamma-absorption methods. Jones (1973) has, very briefly, discussed the sensitivity of neutron systems for the measurement of void fraction. However, for in-pile operation, Ager-Hanssen and Døderlein (1958) and Harms et al (1971) show that voids can be determined by diagnosis of the neutron absorption.

Related to the radiation methods are those in which incident radiation is scattered by, or excites a consequential radiation from, the flowing fluid which is, in turn, detected. Four cases may be mentioned here:

Mean Phase Content (Void Fraction)

(1) Application of X-rays excited by β-sources (Miropolsy and Shneyerova (1962)).
(2) When a channel, containing a two-phase, heavy-water, heavy-water-vapour mixture, is irradiated with gamma rays, neutrons are emitted and can be detected. The rate of emission is in proportion to the amount of heavy water and this fact has been used as a basis for a method of void-fraction determination developed by Rouhani (1962); (see also Rouhani and Becker (1963) and Rouhani (1965) (1966)). The advantage of this method is that the response is direct, rather than being the difference of two large quantities, as is often the case in the gamma-absorption method. However, its range of application must be considered limited, since economics usually forbid the use of heavy water.
(3) Kondic and Hahn (1970) suggested that the measurement of scattered gamma radiation could form the basis of a method for the determination of local void fraction. A beam of gamma rays is passed through the fluid and is scattered in, say, an isotropic manner. The scattered radiation enters a detector through a collimator, so that the intersection of the incident beam and the "detection beam" gives a highly localised measurement volume. Kondic and Hahn show that by measuring the total beam attenuation, as well as the scattering, and by choosing the scattering angle, the scattered intensity can be interpreted to give the local void fraction. The application of this technique to local void fraction measurement in rod bundles is described by Zielke et al (1975) and recent developments of the technique are described by Kondic and Lassahn (1978). Very long counting-times are required (of the order of 1 hour or more) even with primary sources of 25 curie strength. Problems may exist in keeping rig conditions steady for this period.
(4) Neutron scattering methods for void fraction measurement have met increasing use in recent years. The use of the technique is described by Rousseau et al (1976), Rousseau and Riegel (1978), Banerjee et al (1978a) and Banerjee et al (1978b). The section in which void is to be measured is placed in a fast/epithermal neutron beam (from an accelerator target, say) and the scattered and transmitted fluxes are measured by counting. The arrangement is illustrated in Figure 3.17; it is found that if the incident beam is at a relatively uniform intensity then the scattered thermal flux will be dependent on the amount of hydrogeneous material at the cross section and not on its distribution.
Banerjee et al conducted void distribution tests using aluminium/water models as illustrated in Figure 3.18-(A). As is seen from Figure 3.18(B), the expected independence from void orientation was observed.

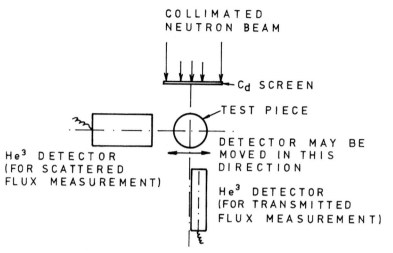

Figure 3.17: Neutron scattering method for void fraction (Banerjee et al (1978b)).

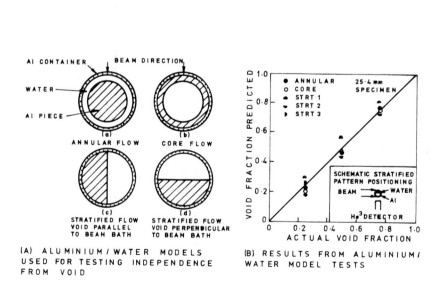

Figure 3.18: Testing of void orientation independence of the neutron scattering method for void fraction (Banerjee et al (1978b)).

Mean Phase Content (Void Fraction)

Impedance Methods

The second most widely used class of methods for void fraction are those involving the measurement of conductance and capacitance. The electrical impedance of a two-phase flow depends on the concentration of the phases and the use of impedance measurement is attractive for void fraction determination, since it gives virtually instantaneous response. Early trials with this technique were not encouraging but, with improved electrode designs, better results have been obtained. Depending on the system, the impedance will be governed by conductance or capacitance or both. Satisfactory results, with impedance gauges, have been obtained by Hoogendoorn (1959), Killian and Simpson (1959), Orbeck (1962), Cimorelli et al (1965), Spigt (1966), Deruaz (1967), Schenk (1967), Cimorelli and Evangelisti (1967) (1969), Milliott et al (1967), Cimorelli et al (1968), Karizawa and Sakurai 1968), van Vonderen and van Vlaardingen (1970), Cybula (1971), Carrard and Ledwidge (1971), Herzberger and Bonvini (1973), Davies and Unger (1973), Subbotin et al (1974b), Herzberger and Hufschmidt (1974), Maitra and Raju (1975), Rosehart et al (1975 and Merilo et al (1977).

The relationship between void fraction, α, and the admittance, A (reciprocal of impedance) will depend on pattern. For fine particles in a continuum, the relationship follows the classical Maxwell (1881) equations. Thus, for a dispersion of gas bubbles in a liquid, we have:

$$\alpha = \frac{A - A_c}{A + 2A_c} \cdot \frac{\varepsilon_G + 2\varepsilon_L}{\varepsilon_G - \varepsilon_L} \tag{3.17}$$

where A_c is the admittance of the gauge when immersed in the liquid phase alone and ε_G and ε_L are the gas phase and liquid phase *conductivities* if the conductivity is dominating and the *dielectric constants* of the gas and liquid if the capacity is dominating. For liquid droplets dispersed in a gas, the Maxwell equation gives:

$$\alpha = 1 - \frac{(A\varepsilon_L - A_c\varepsilon_G)}{(A\varepsilon_L + 2A_c\varepsilon_G)} \cdot \frac{(\varepsilon_L + 2\varepsilon_G)}{(\varepsilon_L - \varepsilon_G)} \tag{3.18}$$

If impedance gauges are operated in conditions in which the conductivity is dominating, drifts occur due to changes in liquid conductivity. This can be avoided by operating at a sufficiently high frequency to give domination by capacity.

The major difficulty with the impedance gauge is the sensitivity to flow pattern. Clearly, for given value of

admittance ratio (A/A_0), equations 3.17 and 3.18 will give very different values of void fraction. Also, the relationship between A/A_0 and α is different if the flow is separated rather than dispersed. Calculations to illustrate this point are reported by Bouman et al (1974) and are illustrated in Figure 3.19. The apparent void fraction, corresponding to a given admittance ratio, would be expected to vary over a very wide range, depending on the flow pattern. This fact must make the method somewhat suspect, though excellent results are reported for it. For instance, Spigt (1966) compared void fractions measured using an impedance meter with those measured using gamma absorption and the comparisons are illustrated in Figure 3.20. However, the work of Van Vonderen and Van Vlaardingen (1970) would seem to indicate that variations do occur. An extensive investigation of the detailed design and performance of the impedance method is that of Olsen (1967). He studied a number of alternative electrode designs (eg those illustrated in Figure 3.21) and suggests that, for measurements in bubble flows, the void meter should have the following properties:
(1) Homogeneous field between electrodes.
(2) Several bubble diameters between sources of current.
(3) Uninterrupted channel cross-section.
It is impossible to meet all these requirements, particularly if the cross-section of the channel is cylindrical. Thus, it is necessary to calibrate the instruments under conditions close to the actual measurement conditions, if trustworthy results are to be obtained.

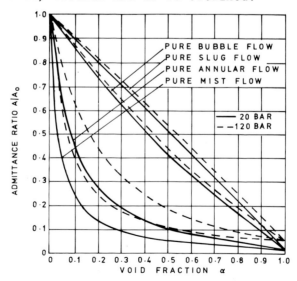

Figure 3.19: Effect of flow pattern on conductivity in the conductance method for void-fraction. Curves calculated by Bouman et al (1974)).

Mean Phase Content (Void Fraction)

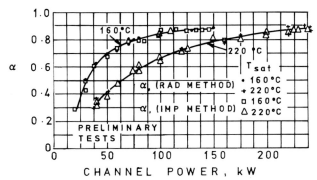

Figure 3.20: Comparison of impedance and gamma void-gauges (Data of Spigt (1966)).

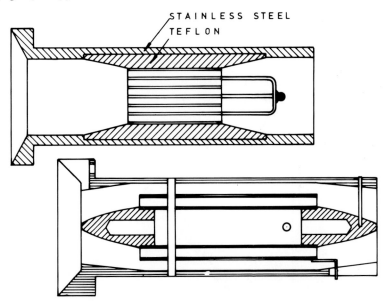

Figure 3.21: Electrode types for impedance void meter studies (Olsen (1967)).

A recent development of the impedance technique by Merilo et al (1977) appears to offer some promise in overcoming at least some of the above difficulties. The principle of their method is illustrated in Figure 3.22. Six electrodes are mounted flush with the channel wall and respective pairs of these are energised by oscillators such that the electric fluid vector rotates as illustrated. By taking an average of the three pairs, a more valid mean void fraction can be obtained. Comparisons between this method and the thick closing valve method are shown in

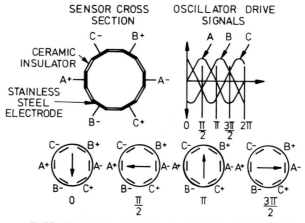

Figure 3.22: Principle of rotating field impedance method for void fraction (Merilo et al (1977)).

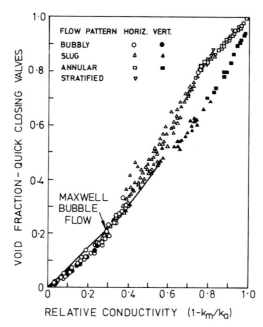

Figure 3.23: Comparison of measurement of void fraction using the quick closing valve and rotating electric field methods (Merilo et al (1977)).

Figure 3.22.

Direct and Indirect Volume Measurement

A third common class of void-fraction measurement technique, is that involving the direct or indirect measurement of the volume of the liquid or vapour phase within the channel.

Direct volume measurement can be achieved by the use of "quick-closing valves". Valves, at the beginning and the end of the section under consideration, are operated quickly and simultaneously. This can be achieved either by mechanical linkage or by the use of electro-magnetically actuated valves. The technique has been used for many years for adiabatic gas-liquid and vapour-liquid flows. Examples of the use of the technique for tubes are given in the papers of Johnson and Abou-Sabe (1952), Hewitt et al (1961), Yamazaki and Shibi (1968), Oliver and Young Hoon (1968), Roumy (1969) and Serizawa (1974). Schraub et al (1969) have used the technique for studies in 9-rod bundles. More recently, developments of the technique made at the CISE Laboratories in Italy (Colombo et al (1967), Agostini et al (1969) and Agostini and Premoli (1971)), have allowed the use of the technique for heated two-phase flows and for transient measurements. Figure 3.24 shows results obtained for a heated channel using this technique. In transient tests, it can be used to isolate sections of the circuit during blow-down. The valves developed in this work have sizes up to $2\frac{1}{2}$ inches diameter and can operate up to 1500 psi and 350°C, with closure times of 10-15ms. To measure the total blow-down transient, repeated re-establishment of the original steady-state conditions is necessary and, typically, 15 or more individual measurements are required in the first 2-3 seconds of a blow-down transient. Reinstatement of the conditions for each new test takes about 15 minutes.

Indirect measurements of void volume depend on the fact that, in a closed circuit, the generation of voids must necessarily be accompanied by some expulsion of the liquid phase from the system. By expressing the liquid into a burette and measuring the volume expelled, one can get a measure of the total voids generated. The success of the technique depends on the design of suitably small condensers in which the vapour generated in the heated section of interest can be condensed to liquid in a small volume. Applications of this method have been described by Kemp et al (1959), Duke and Schrock (1961), Schrock and Selph (1963) and Possa et al (1965). The method is unlikely to be suitable for transient measurements, due to the problem of extraneous volumes outside the section of interest.

Miscellaneous Methods

Other methods which have been used for void fraction

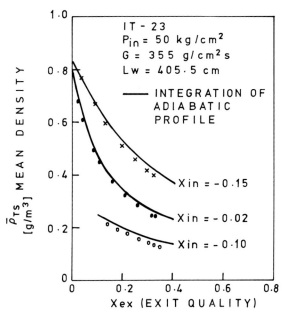

Figure 3.24: Mean density data, obtained by quick-closing valve technique, for heated tubes (Agostini et al (1969)).

measurement includes:
(1) Acoustic techniques. The velocity of sound in a two-phase mixture is highly sensitive to the void fraction and sound velocity measurement has been proposed by a number of authors as a means of determining void fraction (Chedville et al (1969), Gavrilov (1970), Arave (1970a), van der Walle (1965), Technische Hogeschool, Eindhoven (1962), Bonnet and Osborn (1971a) (1971b) and Arave et al (1978). Limitations on the technique are that the sound velocity may depend on the void size, as well as void fraction, and can also vary with sound frequency. Nevertheless, the acoustic technique can sometimes be used where the other techniques mentioned here are unsuitable (Fincke et al (1978)).
(2) Measurements of average phase velocity. The average phase velocity is given by the ratio of phase volumetric-flux per unit area and phase volume fraction. If the volumetric flux is known, and the average velocity determined, the volume fraction can be estimated. One possible method of measuring average velocity is to use radioactive tracer injection and determinations using this technique are described by Evans et al (1971) and Rea and Chojnowski (1974). A further discussion of tracer techniques is given in Chapter 6.
(3) Electro-magnetic flow metering. Again, the principle

of this method is an independent measurement of the average liquid velocity, from which the void fraction can be calculated. Experiments on the technique are reported by Hori et al (1966), Milliott et al (1967), Aladyev et al (1971) and Fitremann (1972). In the experiments of Hori, simulated flow-patterns were set up using plastic models through which the liquid flowed. The liquid velocity was measured and the errors in void fraction determined. It was concluded that the technique could be used, with calibration, over a wide range of flow patterns. This technique is particularly suitable for liquid-metal systems, though it can also be used for air-water and steam-water systems.

(4) Optical methods. An obvious, though tedious, technique is to take photographs of the two-phase flow and to analyse them to measure the total gas volume. Further studies of this type are reported by Griffith et al (1958). Alternatively, the content can be measured using light-scattering (Davis (1955)), though this technique has a very limited range of application.

(5) Microwave absorption. The application of this technique to organic reactor-coolant void measurement is described by Stuchly et al (1973). Another interesting application of microwaves to fluid density (void fraction) measurement is described by Wenger and Smetana (1972). They determined the resonant frequency of a microwave cavity which formed part of the flow line which was carrying liquid hydrogen or hydrogen vapour liquid mixtures. The resonant frequency was related to the mean density. Kraszewski (1971) describes the use of a microwave sensor for water-content measurements in two-phase mixtures such as margarine.

(6) Nuclear magnetic resonance. This method was the subject of a British patent (1968) and has recently been further investigated by Lynch and Segel (1977). When nucleir possessing a spin angular momentum are introduced into a magnetic field they take up specific orientations and, in NMR experiments, the amount of radio-frequency energy required for reorientation is measured. Lynch and Segel report very promising results for this method; in flowing systems, it is important to ensure that the relaxation time for the reorientation is small compared with the residence time in the measurement region. This can be achieved by doping the liquid with a paramagnetic salt (eg copper sulphate if the liquid is water).

(7) Link between pressure and flow oscillations. Hayes (1972) has described a method for measuring the void fraction in sodium systems. Pressure fluctuations, induced by a pump, cause flow fluctuations. The relationship between the pressure fluctuations and the

flow fluctuations depends on the void fraction.
(8) Infra red absorption methods. These show considerable promise, particularly when used at very high void fraction. They can be used to measure the component concentrations in air/steam/water mixtures and a successful application of the methods is described by Barschdorff et al (1976).
(9) Neutron noise analysis. In the cone of nuclear reactors, it is possible to deduce the void fraction from correlation analysis of the signals from neutron detectors. This technique has been used by Ando et al (1975) and Ceelen et al (1976).

A large variety of techniques have been proposed for void fraction measurements, reflecting the great importance of this parameter. For a precise measurement, the quick-closing valve technique appears to show the best results, even for heated channels at high pressure and at transient conditions, though often, it cannot be applied. The gamma-ray absorption technique is often more convenient, but may suffer from inaccuracies due to the unknown distribution of the phases and to the intermittency in the flow. Some of these problems may be minimised by continuous counting and by using strong hard sources. More localised measurements can be obtained using gamma scattering. For rapid time-response, the impedance method would appear to be encouraging, though one would expect it to be sensitive to the flow pattern, in addition to the void fraction. Many techniques are appropriate in specific situations and the designer should always review the full range of possibilities in choosing the method for a given system.

Critical Heat Flux

Obviously, critical heat-flux in the steady state is an important design limitation. For heat-flux controlled systems (electrical and nuclear heating), it is defined as the flux at which a small increase in heat flux gives an inordinate rise in wall temperature. For wall temperature controlled (fluid heated) systems, it is defined as the condition in which a small increase in wall temperature leads to an inordinate reduction in heat transfer coefficient. Normally, the changes in conditions is considerable and the critical condition can be easily established. However, for high-pressure systems, (170 bar, say), the changes are relatively small and this increases the difficulty of detection. Normally, in uniformally-heated channels, burn-out occurs at the end of the test section but in non-uniform heating, the location of burn-out can often be upstream of the end of the heated zone. In considering measurement techniques for burn-out, it is useful to understand the mechanisms, at least qualitatively, discussions of these mechanisms are given by Hewitt and Hall-Taylor (1970), Tong and

Hewitt (1972), Whalley et al (1974) and Hewitt (1978).

Since most experiments on critical heat-flux have been conducted with direct electrical heating systems, the detection of an inordinate temperature rise has been possible by balancing the voltage drop, across that part of the test section where the first onset of the critical conditions is likely, against the voltage drop in a section immediately upstream. When overheating occurs, the bridge becomes off-balance and the off-balance voltage can be used to trigger a warning or, if necessary, to switch off the power input. The design of such systems is discussed by Green (1967), Kichigin (1976) and Jackson (1976). The principle is illustrated in Figure 3.25; the voltage drop across the zone of channel in which the critical phenomenon is likely to occur in V and, in the pre-critical condition, resistances R_3 and R_4 are adjusted such that $R_1/R_2 = R_3/R_4$ and δV, the off balance voltage is zero. Suppose that part of the channel in the region corresponding to R_1, experiences a temperature excursion ΔT. The off-balance voltage is given by:

$$\delta V = \frac{V R_1 \gamma \beta \Delta T}{R_1 + R_2} \qquad (3.19)$$

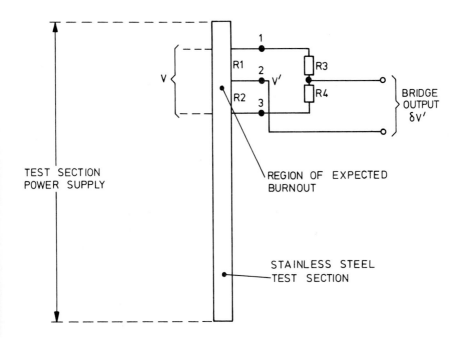

Figure 3.25: Bridge system for critical heat flux detectors.

where γ is the fraction of the surface in the region corresponding to R_1 which overheated, and $(1 + \beta\Delta T)$ is the factor by which the electrical resistance of the overheated zone increases.

The critical condition can be detected using thermocouples (eg Chojnowski and Wilson (1972) and Bailey (1977)) and this technique can also be used for in-pile and indirectly electrically heated systems. Where dryout is occurring on, say, the inside of the tube, the thermocouples can be placed on the outside without disturbing the flow. The same kind of problems on thermocouple installation occur as those reviewed above. If the thermocouples are insulated from the electrically-heated tube wall, then their response time is large. Mayinger et al (1967) show that mica-insulated thermocouples have a response time as long as 280ms. The need for insulation can be avoided by using a 3-wire thermocouple system where the wires are directly attached to the surface and the responses from the respective pairs balanced. This system has a much more rapid response, ie about 7ms. (Mayinger et al (1967)). Installation of internal thermocouples inside rods, when the flow is outside (eg in annuli or rod-bundle experiments) presents more severe difficulties, though it is still possible to achieve reliable results (Moeck et al (1966), Gustafsson and Kjellen (1971)). In the UKAEA, methods have been developed for the detection of dry-out in rod clusters which do not depend on the attachment of thermocouples to the inside of the cluster. A detailed description of these methods is given by Adnams et al (1972). The devices used in this work consisted of plugs which could be inserted down the centre of the electrically-heated tubes. These plugs carried temperature detectors which could either be thermocouples, radiant thermopiles or induction-element detectors. A considerable amount of development work has been done on these devices and the sort of complexity necessary is illustrated in Figure 3.26.

For nuclear or indirectly-heated rods, thermocouples must be attached to the outside of the rod in order to detect burn-out. Since burn-out often depends on the breakdown of a relatively thin liquid film, and since the thermocouple itself can act as a point of capture and re-distribution of liquid onto the surface, it must be considered generally unsatisfactory, and unrepresentative of the real situation, to have thermocouples simply attached to the outside of the rods. This problem can be overcome by using grooved rods and inserting the thermocouples in the grooves, making good the surface by brazing. This method is discussed by Wikhammer et al (1964) and has been used extensively in experiments at Chalk River, in both out-of-pile and in-pile loops. It is extremely difficult to obtain an absolutely uniform surface over the thermocouples but, at least, it is better than having the

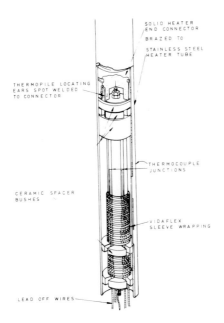

Figure 3.26: Radiant thermopile burn-out detector (Adnams et al (1972)).

thermocouple itself exposed on the outside of the tube. Alternatively, a double sheath can be employed with the thermocouple located in a depression in one of the sheaths (Clark et al (1974)).

In some circumstances, the dryout occurrence can be directly observed and this, in itself, is a good method of detecting dryout. Some observations have been made by Hewitt et al (1965) and by Parsons and Brundrett (1971). The technique can only readily be applied to the observation of burn-out on the outside of rods and suitable windows have to be introduced to allow viewing of the surface. Thus, the technique has very limited application.

An alternative to thermocouple and resistance detection for burn-out on the outside of tubes, particularly suitable where the burn-out is likely to be highly localised and where the temperature coefficient of resistivity of the material is low, is the use of infra-red burn-out detection. A system for such detection has been developed by Benn and Shock (1974). However, this method can only be applied when a satisfactory view factor can be obtained of the heated surface.

Measurement of the point of onset of the critical phenomenon in fluid heated system gave rise to more difficult problems. Critical heat flux measurements for freon flowing in liquid metal heated tubes are described by France (1974). Here, the axial temperature distribution of the flowing liquid metal heating fluid was determined and, by differentiation of this profile, the local heat flux was determined. The point at which this heat flux reached a maximum was identified as that corresponding to the onset of the critical phenomenon.

For electrically-heated systems, the bridge-balancing and thermocouple techniques are well established for the detection of critical heat-flux. For more complex geometries, such as rod bundles, the systems of critical heat-flux detection are often complex. However, even in these cases, satisfactory operation is feasible using existing techniques. For nuclear heated systems these methods are inapplicable. For some applications, the infra-red detection technique would appear to justify further development. Also, a general method is needed for application to nuclear-fuel heat-transfer experiments and more studies are required in fluid heated systems.

4 Primary Design Parameters (Fault Conditions)

Introduction

Experiments on two-phase flows in the context of nuclear reactor safety are now, of course, very widespread and the performance of these experiments involves proper attention to detail on two-phase instrumentation. In these experiments, a whole spectrum of techniques has been employed, and will be employed in the future, and these technqiues are covered in the various sections of this report. In other words, with many techniques, there is nothing specific in the fact that the experiment is being done in the context of nuclear reactor safety and the contents of all of this report can be used as appropriate. However, safety experiments have thrown up a number of features which demand specialised experimental techniques and, in this chapter, the following examples of such features are discussed:
(a) System discharge and related parameters.
(b) Bubble growth and collapse.
(c) Dryout under fault conditions.
(d) Rewetting of hot surfaces.

A survey of the general problems of theoretical and experimental work on accident analysis is given by Boure (1975) and Griffith (1978). A useful survey of measurement techniques is also given in Collier (1975); the report of Brockett and Johnson (1976) is aimed mainly at this area of experimental study.

System Discharge and Related Parameters

Of prime importance in considering the safety of reactors is the rate of discharge of the reactor contents through circuit and vessel ruptures. The rate of discharge is limited by critical flow and it is worthwhile mentioning briefly some of the background to measurements of critical flow in two-phase systems.

Since two-phase flows often have a high compressibility coupled with a high density, relative to a gas, the critical velocities are rather low. In general, the critical velocity passes through a minimum as the quality goes from zero (all liquid), to unity (all vapour). With discharge from a liquid-filled system, one difficulty is calculating the actual exit quality, since this can be much lower than the thermodynamic equilibrium quality (see, for instance, Smith et al (1964)). A major experimental difficulty, in studying two-phase critical flow, is that of measuring the exact value of the pressure at the point of discharge. Immediately upstream of this point, the pressure gradient is extremely high and it is difficult to place pressure tappings exactly at the end of the tube. A method of overcoming this, for critical flow in a venturi, is reported by Smith (1971a) and is illustrated in Figure 4.1. The venturi was made in an annular form with the centre shaped portion moveable with respect to a fixed pressure tapping on the outside wall. Accurate and reproducible pressure measurements, right through the critical point, were possible by this method. The problem with measuring the outlet critical pressure is also discussed by Takekoshi et al (1969), who used pressure tappings within the plate of the discharge orifice in their experiments. Other recent experimental studies of critical two phase flow include those of Seynhaeve (1977) and Prisco et al (1977).

Critical flow velocity and sonic velocity are usually almost identical in single-phase flows. However, in two-phase flows, the identity is not so clear and much work has been done on sound-wave, and pressure-shock propagation

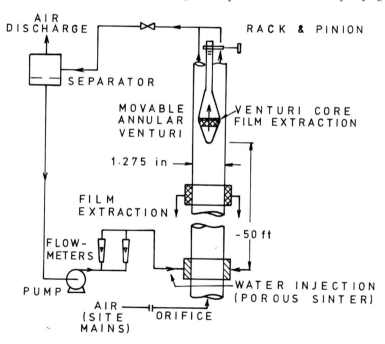

Figure 4.1: Annular venturi used for precise pressure-measurement in critical two-phase flow (Smith (1971a).

in two-phase flows. Transmission of sound from a sound
transducer has been investigated, for instance, by Mecredy
et al (1970), Kokernak and Feldman (1972), Mukhachev and
Susekov (1972), Kielland (1967), Feldman et al (1971) and
Alforque et al (1977). Kokernak and Feldman made measurements on two-phase vapour-liquid bubbly flow of refrigerant
-12. The sound velocity was measured in a direction normal
to the direction of flow and was determined as a function
of frequency. The results are illustrated in Figure 4.2.
At low frequency, the sound velocity is strongly dependent
on two-phase quality but, at high frequency, the sound
velocity is equal to that in a liquid phase alone. In the
intermediate range of frequencies, there is a resonance
phenomenon in bubbly flow, as the result of which, the
velocity of sound shows a sharp minimum. A theoretical
model for sound propagation in two phase flow (including
non-equilibrium effects) is reported by Ardon and Duffey
(1977).

Many experiments have been reported on the propagation
of shock waves, (pressure pulses), through two-phase flows.
These studies include those of Karplus (1961), Semenov and
Kosterin (1964), Collingham and Firey (1963), Nyer et al
(1967), Hamilton et al (1967), Eddington (1967), Evans et
al (1970), Bockh and Chawla (1972) and Padmanabhan (1976).
The propagation of the shock wave is determined by introducing pressure transducers at various distances along the
channel and measuring the time delay of the passage of the
shock from one transducer to the next. The shock wave may
be considered as being composed of very high-amplitude
sound waves in a Fourier spectrum, so designed as to
represent the actual shape of the shock. The transmission
of sound in two-phase flow is, as was shown in Figure 4.1,
a function of the frequency and the transmission of this

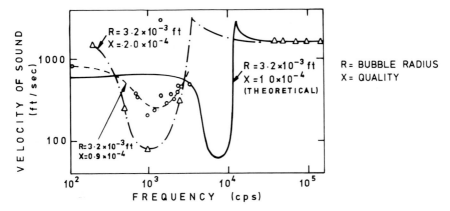

Figure 4.2: Influence of sound frequency on sound velocity, for
transmission through a two-phase bubbly flow (Kokernak and Feldman
(1972)).

shock will thus be somewhat complex. However, the high
frequency components are usually quite small and the shock
transmission occurs at a well-defined velocity in most
experiments. Nyer et al (1967) show that the propagation
is fairly well represented, theoretically, by assuming the
two-phase mixture to be homogeneous. An interesting
experiment on shock propagation, in a static system, is
reported by Bockh and Chawla (1972) and is illustrated in
Figure 4.3. The transmission of a pressure pulse in a tube
containing a static layer of water was investigated and the
pulse-transmission velocity measured as a function of the
water content. It was found that the principal part of the
pressure change was transmitted at the low velocities
associated with a two-phase mixture, thus demonstrating
that, even in this extreme case, gas-liquid interaction is
sufficient to give the expected reduction in transmission
speed.

   Although a fundamental knowledge of critical flow is of
value in assessing discharge rates from ruptures, it is
still necessary to make direct studies of system blow-down.
An increasing body of data is becoming available in this
area. Experiments of this category include those of
Carzaniga et al (1969), Uchida and Shimamune (1969),
Borgartz et al (1969), Rudiger (1972), Tokumitsu (1972),
Shimanune et al (1972), Rasmussen et al (1974), Adachi
(1974), Hayashi (1974), Aerojet Nuclear Company (1974a)
(1974b), Flinta (1975), Appelt and Kadlec (1975), Frix et
al (1975), Henry (1975), Premoli and Hancox (1976),
Simpson et al (1977), Edwards et al (1977), Arrison et al
(1977), Narial (1977), Marshall and Holland (1977), Ulber
and Lubesmeyer (1977), Fisenko and Sychikov (1977),
McPherson (1977) and Rousseau and Riegel (1978). Experi-

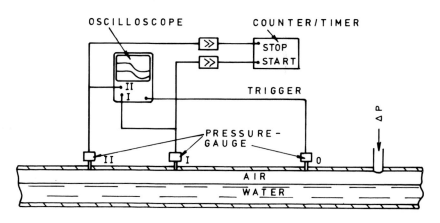

Figure 4.3: Sound transmission measurements in a tube containing a
static layer of water (Bockh and Chawla (1972)).

ments range from those on a single tube, as exemplified by the work of Borgartz et al (Figure 4.4), through to large-scale tests on simulated reactor vessels (eg Rudiger (1972) - Figure 4.5) and to the large scale loss-of-coolant tests being carried out in the LOFT nuclear experiments in the USA (McPherson (1977)). Many of tne measurements required in these experiments have been reviewed elsewhere in this report. Some specific problems are worthy of mention, however, and are as follows:

(1) Transient mass-flux measurement. Obviously, the amount of fluid discharged from the system is of prime importance in considering fault design. The most straightforward way of measuring the outlet flow-rate is by introducing an orifice or another flow-meter in the system, in a region which is known to be single-phase, (for two-phase flow, the orifice reading depends on the quality as well as the mass velocity). This technique was used by Flinta (1975). Alternatively, the whole system can simply be mounted on a load cell and the change of weight with time determined (Rudiger (1972)). Aerojet Nuclear Company (1974) are developing cross-correlation methods for velocity measurement, using thermal pulses. Further discussion of steady state and transient mass flow measurement is given in Chapters 5 and 6.

(2) Transient void-fraction measurement. Probably the most convenient way of measuring void fraction, in rapidly changing systems, is by using X-ray absorption. This technique was discussed above and has been employed for blow-down experiments by Borgartz et al (1969); the equipment used by these authors is shown schematically in Figure 4.6. Recently, high-intensity gamma sources (up to 25 curies) have been employed to obtain time resolution of the order of 1ms. Westinghouse

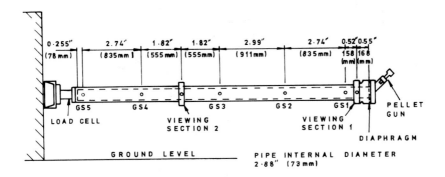

Figure 4.4: Apparatus for blow-down studies in a single tube (Borgartz et al (1969)).

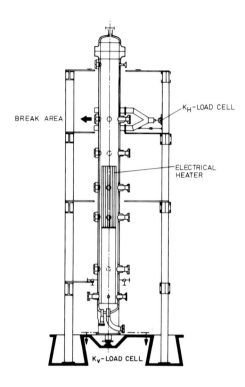

Figure 4.5: Apparatus for blow-down studies from a vessel simulating a reactor system (Rudiger (1972)).

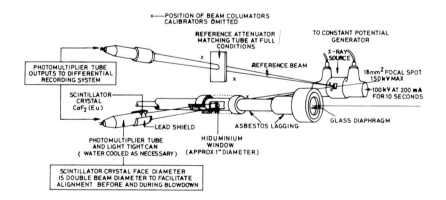

Figure 4.6: Double-beam X-ray system for density measurements in blow-down system (Borgartz et al (1969)).

Electric Corporation (1975) estimated void distribution in a simulated fuel element, during blow-down, by making accurate transient pressure-distribution measurement. They deduced the void fraction from the static-head change. Such a system could only be used where it was certain that the frictional and accelerational component of pressure gradient were either negligible or could be calculated accurately.

(3) Outlet momentum flux. This can be conveniently measured by using a load cell, operating in a direction opposite to the discharge, as illustrated, for example in Figure 4.5.

(4) Local pressure transients. Here, fast-response pressure transducers are required, for example, piezo-electric transducers. Borgartz et al (1969) discuss, in detail, the problems of using such transducers. Unless great care is taken in providing adequate cooling of the transducer and in obtaining the correct dynamic response, then confusing results can be obtained. Borgartz et al found the best dynamic response was obtained by using a silicone rubber plug, 1/16 inch thick in from of the transducer. Frequent calibration of the pressure transducers is recommended.

(5) Rupture mechanism. It is sometimes difficult to obtain reproducible results for a simulated rupture. A typical arrangement for bursting a rupture disc is illustrated in Figure 4.7 (Hayashi (1974)).

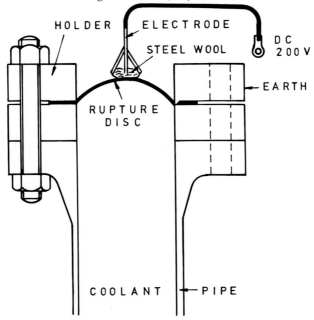

Figure 4.7: Rupture mechanism for blow-down studies (Hayashi (1974)).

(6) Fuel-pin simulation. A great deal of work has been carried out on the development of indirectly-electrically heated pins for fuel simulation for blow-down experiments. Typical of this work is that reported by Uchida and Shimamune (1969). Pins for such experiments are now becoming commercially available, but still remain very expensive, particularly if a non-uniform heat flux is required.
(7) Nuclear radiation effect. In-pile experiments often lead to sensors and cables being exposed to steady and transient radiation fluxes. These effects have been investigated by Terry et al (1965) and Langdon et al (1970).

The understanding of critical discharge phenomena is still incomplete, particularly in terms of local parameters of the flow. For blow-down measurements, there is much scope for further development work. Much further work is necessary on the development of existing and new methods for measuring outlet mass flux, void fraction and phase enthalpies.

Bubble Growth and Collapse

Bubble growth and collapse phenomena are important in a number of contexts including:
(1) In the detection of void formation within a reactor system.
(2) In the study of pressure-and void-transients, associated with rapidly-changing heat fluxes or pressures.
(3) In situations leading to "vapour explosion" due to sudden vapour growth following a core meltdown in, say, a sodium cooled fast reactor.

Void detection in sodium-cooled nuclear reactors is important, since the reduction of flow, as a result of a flow blockage of one of the channels, gives rise to an increased temperature which, in turn, gives rise to nucleate boiling. Characteristically, nucleate boiling emits noise and study of boiling noise, particularly in the context of reactor void-detection, has been widespread in recent years. Papers dealing with this subject include those of Osborne and Holland (1947), Firstbenberg (1960), Walton (1963), Misch (1963), Kartluke et al (1965), James (1965), Macleod (1966), Jordan (1966), Saxe (1966), Kutukcuoglu (1967), Saxe and Lau (1968), Saxe (1969), Hayes (1969), Grolmes and Fauske (1969), Ponter and Haigh (1969), Anderson and Grate (1969), Ram and McKnight (1970), Anderson et al (1970), Kichigin (1970), Bosio et al (1970), Schmidt et al (1970), Saxe et al (1971), Haigh and Ponter (1971), Gross et al (1971), Bonnet and Osborn (1971a). Bonnet (1971), Bonnet and Osborn (1971b), Lykov (1972), Winterton (1973), Tolubinsky et al (1974), Robinson et al (1974), Burton (1974), Carey and Albrecht (1976), Kikuchi et al (1977), Hanus et al (1977), Benkert et al (1977) and Catling et al (1977). Other methods of detecting

nucleate boiling in reactors include the use of ion
chambers (Boyd (1959)), and measurements of single-bubble
growth-rate in liquid sodium, by ultra-sonic techniques,
are discussed by Kazemeini and Ralph (1975). Single-bubble
growth measurements in sodium were carried out by Singer
and Holtz (1970) using electrical detection methods.
Liquid sodium flow boiling studies on multi-rod geometries,
whose objective was to examine the boiling process following a flow reduction due to blockage, are reported by
Brook and Peppler (1977) and by Kaiser and Peppler (1977).

The first initiation of boiling in the subcooled region
is often very difficult to detect from measurements of
wall temperature or void fractions. Nevstruyeva et al
(1973) show that a sensitive indication of the first
nucleation points can be obtained by a salt deposition
method. The incipient boiling onset can be anticipated by
using induction of cavitation on an ultrasonic probe
(De Prisco et al (1962). Kutukcuoglu et al (1967) report
determination of subcooled boiling onset by measuring
heater rod vibrations.

Photographic methods have been widely used in studying
bubble growth and collapse, in transient conditions. Lurie
and Johnson (1962) and Schrock et al (1966), used high-
speed cine-photography (4200 frames per second) in studying the transient pool boiling of water with a step in
heat generation. Board (1969) describes bubble growth
studies in multipin geometries, using photographic
techniques. Grolmes and Fauske (1974) reported the use of
high-speed photography in studying the actual propagation
of free-surface boiling in super-heated liquids. High-
speed photographic studies of bubble collapse have been
reported by Suezawa et al (1972), Green and Mesler (1970),
Kudo (1974), Adlam and Dullforce (1974), and Pitts et al
(1977). In the work of Green and Mesler (1970), high-
speed photography, up to 35000 frames per second, was
employed. Bubble growth studies using fast, coaxial thermocouples, are described by Schmidt (1977).

Transient boiling and condensation is often accompanied
by the generation of pressure waves, (the small pressure-
waves generated by bubble growth and collapse in nucleate
boiling, are the source of the boiling noise used in
reactor void-detection, as described above). Studies of
pressure transients, in boiling and condensation (including
cavitation, are reported by Mellen (1954), Ellis (1966),
Uchida (1969), Green and Mesler (1970), Suezawa et al
(1972), Adlam and Dullforce (1974) and Grolmes and Fauske
(1974). A specific example of the generation of pressure
waves by rapid condensation is the so-called "waterhammer" effect. In a pressurised water reactor (PWR), this
can occur during abnormal operating transients where a
cold water slug is introduced and traps a steam void which
rapidly condenses and causes the water to accelerate rapidly; when the water slug hits an elbow, a pressure wave is

generated which propagate through the piping system. Studies of this phenomenon are reported by Rothe (1977).

The direct measurement of void growth during transient boiling, using X-ray absorption, has been reported by a number of authors (Schrock and Selph (1963), Johnson (1970), Staub et al (1967), Wentz et al (1968), Jones (1970) and Smith (1971b)). For rapid transients, very high intensity X-ray sources are required and such sources have been used in the study of transient pool boiling, with time constants down to a few milli-seconds (Schrock and Selph (1963) and Johnson (1970)). An alternative method of studying phase growth is to use ultrasonic detection. Bubble growth studies, using a pulsed ultrasonic system, are described by Poole (1970).

Consideration of the very unlikely possibility of a core melt-down in a fast reactor leads to the postulation of a variety of fascinating two-phase flow problems. These are described, for instance, by Fauske (1976). "Vapour explosions" are one form of fuel/coolant interaction which is possible in these circumstances and result from the rapid mixing of a hot liquid and a cold varporisable liquid (ie molten fuel and sodium). Such vapour explosions are also possible in other situations such as the accidental contact of molten steel and water and the ingress of water into pools of cryogenic fluids. The explosion results from the release of a significant fraction of the available energy in the form of a pressure shock wave. There is some dispute about the precise mechanisms involved (Bankoff (1978)), but this field is one of great importance and one in which sophisticated experimental techniques will have to be developed.

A very large number of papers have been published on the acoustic detection of voids in fast reactor systems. Bubble growth and collapse is important in studying reactor transients and cavitation and there are many papers in this area. Satisfactory, high response time, X-ray measuring techniques have been developed. There would appear to be scope for the development of improved X-ray methods or other methods, for measurement of bubble collapse in liquid metal systems.

Dryout Under Fault Conditions

Experimental studies of the onset of dryout, and of temperatures in the post-dryout region, during the blow-down following a rupture, have been reported by Hicken et al (1972), Morgan et al (1972), Friz and Riebold (1972), Class et al (1972), and Imhoff and Murray (1972). In the blow-down situation, the behaviour depends on the changes with time in the pressure, the mass-flow and (in both the reactor and in some experimental simulations) the heat flux. This combination of factors makes the interpretation of blow-down data very difficult and better understanding of the phenomena can be obtained by carrying

out experiments in which only one of the variables is changed with time. Examples of this kind of approach are as follows:
(1) Dryout with transient pressure changes. For a pool-boiling situation, Shimamune et al (1969) studied the temperature transients in a heated metal foil suspended in water in a pressure vessel. The type of transient behaviour observed, depended on the initial conditions (sub-cooled water or two-phase). Figure 4.8 shows one form of behaviour (observed with water sub-cooled before the depressurisation) in which large wall-temperature oscillations are observed.
(2) Heat input transient. Studies of transient dryout at constant flow, with a power surge, are reported by Gaspari et al (1971) (1972) and Aoki et al (1974). The power rose to its final value over periods of 0.5-1 seconds.
(3) Flow transients. Dryout with flow transients, at constant pressure, was studied by Gaspari et al (1971) (1972), Moxon and Edwards (1967), Shih (1974), Gaspari and Granzini (1976), Smirnov et al (1977) and Henry and Leung (1977). In the experiments of Gaspari et al, the flow stoppage was almost instantaneous, whereas, in the experiments of Moxon and Edwards, it

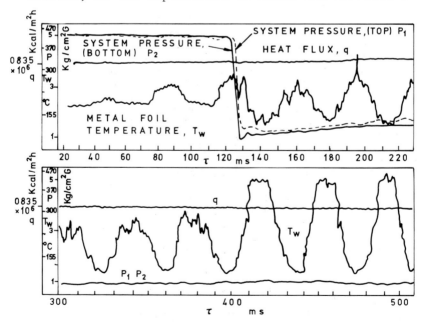

Figure 4.8: Temperature oscillations of a heated foil, in a static tank of liquid, following sudden depressurisation (Shimamune et al (1969)).

occurred over a period of about 1 second. Gaspari and Granzini (1976) studied transient dryout in rod clusters and Henry and Leung (1977) used Freon-11 in modeling transients for water systems.

Simultaneous depressurisation and flow transients, for flowing systems, have been studied by Gaspari et al (1971) (1972), Aoki et al (1974), Shih (1974), Belda (1975) and Leung et al (1975). The last-named work is particularly interesting, since it concerns an important limiting case: dryout during a transient flow reversal in the channel.

In studying transient-dryout and post-dryout temperatures, the following main experimental points need to be taken into consideration:

(1) Temperature measurement. The points made above, with respect to the attachment of thermocouples, are very important in transient measurements. Any results obtained with thermocouples, attached to the surface and protruding from it, must be regarded with some suspicion, since these thermocouples can affect the local flow-pattern near the surface. Typically, they could cause an increase in surface temperature, by disrupting any liquid film present, or they could collect water droplets from the vapour stream and redeposit them on the surface, causing a reduction of surface temperature. No general rules can be given about what would happen, hence the uncertainty. If possible, the thermocouples should be set into grooves in the surface. The use of imbedded thermcouples for indirect heating systems is discussed in detail by Tschuke and Moller (1977) and Engler and Von Holzer (1977). Figure 4.10 shows a typical indirect heater assembly used in emergency core cooling studies.

(2) In transient conditions, dryout may occur in unexpected places. Instrumentation of the end of the channel (the normal location of dryout) may fail to indicate its occurrence in, say, the entrance region.

(3) The behaviour of surface heat-flux, following dryout, may depend on the type of heating used. This is illustrated in Figure 4.9, whic shows the differences in behaviour for directly- and indirectly-heated rods. Systems like the multiple block test section shown in Figure 3.10 can sometimes be used to advantage. Care should also be taken that the occurrence of dryout does not lead to an axial redistribution of heat flux in joule-heated systems.

(4) The experimentalist should always be aware of the potential differences between direct joule, indirect joule and nuclear heating. The important point here is the release rate of stored energy within the element.

An important case of transient dryout, where dryout is desirable rather than an undesirable feature, is that of the onset of film boiling on the walls of the pressure

Dryout Under Fault Conditions

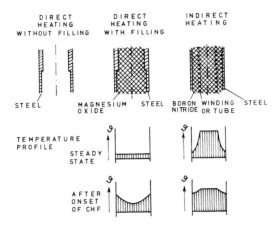

Figure 4.9: Temperature response of heaters of various designs (Hicken et al (1972)).

vessel, following a depressurisation. Too rapid heat liquid, can lead to undesirable levels of thermal stress. This problem has been studied by Hufschmidt and Burck (1972).

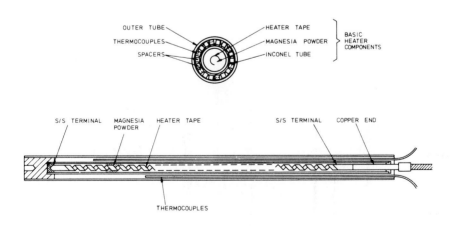

Figure 4.10: Heater rod assembly used in emergency cone cooling studies (Hawes (1976)).

Further development of instrumentation is urgently required for the post-dryout region. This includes measurement of local temperatures, the characteristics of the droplets, wall-temperature fluctuations etc.

Rewetting

Obviously, once dryout occurs within a reactor system, and the temperature rises, then it is important to rewet the fuel element surfaces by some form of emergency cooling system. These systems fall into three main classes:
(1) Top flooding. Here, liquid is introduced from spray nozzles at the top of the fuel element and flows downwards through it.
(2) Bottom reflooding. Here, liquid is fed to the bottom of the channel and refloods the fuel element, a rewetting front gradually passing up the channel.
(3) Spray cooling. Here, water is sprayed in the interstices of the core through one of the rods of the bundle and out through small holes.

The three methods are shown schematically in Figure 4.11.

Studies of bottom reflooding, for both single rods and rod bundles, are reported in the literature (Andreoni and Courtaud (1972), Martini and Premoli (1972), Riedle and and Winkler (1972), Campanile and Pozzi (1972), Hayashi (1974). Griffith et al (1975) and White and Duffey (1975), Waring and Hochreiter (1977) and Lee et al (1978) for example). Typical examples of experiments with top reflooding (film rewetting), are those of Shires, Pickering and Blacker (1964), Bennett et al (1966), Elliott and Rose

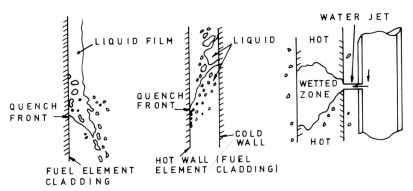

Figure 4.11: Types of reflooding.

(1971), Duffey and Porthouse (1973), and Eriksson (1975). Experiments on spray cooling are reported by Kremnev et al (1970) and Duffey and Porthouse (1972).

Considerable light can be thrown on the mechanism of cooling and rewetting by sprays by considering the interaction of single drops with hot surfaces. Typical of these studies are those of Hein and Liebert (1976), Styrikovich et al (1978) and Nishio and Hirata (1978) for solid surfaces and Waldram et al (1976) for drops impinging on hot liquid surfaces. In all emergency core-cooling systems, the problems of introducing the water into the core are critical and cooling water/steam interactions, in the inlet piping, have been discussed by Lowe (1975). A comprehensive review of both bottom- and top-reflooding has been prepared by Butterworth and Owen (1975) and the mechanisms of reflooding are reviewed by Elias and Yadigaroglu (1976) and Kirchner (1976).

In carrying out experiments on rewetting, the following main experimental points should be borne in mind:
(1) Rewetting of surfaces can be promoted by surface protrusions. An example of the effect of supports on rewetting is shown in Figure 4.12, which is taken from the report of Era et al (1966). Rewetting of the heated centre-rod in an annulus was promoted by support spacers as shown.
(2) The presence of more than one component in the system

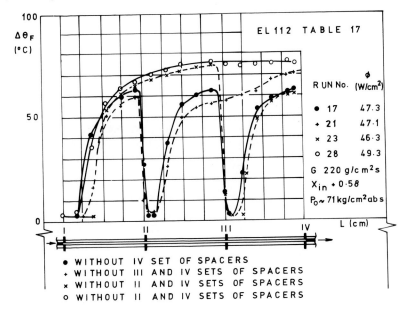

Figure 4.12: Effect of spacers on rewetting, in the post-dryout region in an annulus (Era et al (1966)).

can lead to interfacial temperature differences which, in turn, produce interfacial surface-tension gradients. These gradients create film flows (the Marangoni effect), which can ultimately lead to film breakdown. Discussion of film breakdown, due to Marangoni effects, is given by Ponter et al (1967) and also by Shires et al (1964). In the latter work, it was shown that the film breakdown induction does not occur for a single component (vapour-liquid) system. However, if a second component (such as air) is present, then Marangoni effects are always possible.

(3) Surface condition. Rewetting depends critically on the heat-transfer coefficient on the wet side, immediately upstream of the wetting front. Here, nucleate-boiling heat-transfer is present and the number of nucleation sites directly influences the transient coefficient. Thus, Piggott and Porthouse (1973) have shown that shot-blasting of the surface can cause the wetting velocity to increase by a factor of 3, while silver-plating the surface halved the velocity. The influence of surface deposits is complex; experiments are reported which show both an increase (Piggott and Porthouse (1973) and a decrease (Bennett et al (1966) in wetting rate. This diversity of result probably arises from the fact that oxide deposits, of a porous nature, can actually increase the heat-transfer coefficient, by increasing the number of available nucleation sites, whereas non-porous deposits provide an insulating layer on the surface, coupled with the possible reduction of the number of active sites.

(4) Coupled effects of water flow-rate and sub-cooling on quench-front velocity. Under saturated conditions, and in top reflooding, the rewetting rate is almost independent of rewetting film flow-rate. However, in the presence of cooling, there is an effect of flow rate (Duffey and Porthouse (1973)). Also, the effect of *inlet* sub-cooling will vary, depending on the actual position of the wetting front, since the film has been heated up to that point. In other words, care should be taken over the setting-up of experiments, so as to simulate the actual reactor case. Reproducible results, of a generally applicable nature, can only be obtained by using saturated conditions.

Rewetting of surfaces is affected by a variety of factors including surface protrusions, mass transfer effects leading to Marangoni phenomena, surface conditions (eg oxide deposits) and complex effects of sub-cooling. Further studies are required of the effect of surface conditions on rewetting and of the interactions between droplets and hot surfaces. The last area, particularly, needs the development of better experimental techniques.

5 Secondary Design Parameters

Introduction

Secondary design parameters are ones which need to be investigated as part of the overall design but which are not necessarily the first consideration of the designer. The following items are discussed under this heading in this chapter.
(a) Vibration and transient momentum flux.
(b) Flow distribution.
(c) Stability.
(d) Quality and mass flow measurement.
(e) Liquid level detection.

Vibration (and Transient Momentum Flux)

Vibration problems in both non-nuclear fluid systems have tended to assume greater importance in recent years. Due to the increase of shell-side velocities in heat exchangers, the incidence of failure of heat exchangers due to tube vibration has increased. The mechanisms by which vibration occurs in heat exchangers in cross flow include:
(1) Vortex shedding. Vortices are shed in flow across the tubes and this gives rise to fluctuating forces on the tube; if the frequency of vortex shedding is similar to the natural frequency of the tube, the tube vibrates and this can cause damage due to impaction with other tubes or with the tube supports (eg with baffles).
(2) Turbulent buffeting. Here, within a tube bundle, random flow irregularities may have a dominant frequency coincident with the tube natural frequency.
(3) Fluidelastic instability. Here, vibratory motion of one tube causes variation in the forces on adjacent tubes due to changes in flow configuration. Once initiated, such vibrations are self exiting and growing continuously with increasing fluid velocity. The fluid velocity for the onset of this instability is often calculated by the Connors (1970) equation:

$$V_g/fd = K\sqrt{m\delta/\rho d^2} \qquad (5.1)$$

where V_g is the velocity through the gap between the tubes, f the tube natural frequency, d the tube diameter, ρ the fluid density, m the virtual mass of the tube and δ the logarithmic decrement for damping. K is a constant which varies with the tube configurations.

In axial flow, flow induced vibration may result from fluid-elastic instabilities or from boundary layer pressure fluctuations induced by the fluid. Comprehensive reviews of tube vibration phenomena are presented by Pettigrew (1977) and Shin and Wambsganss (1977). With two-phase flows, the fluctuations which occur in the void content of the flow, make the susceptibility to vibration even worse than in single-phase flow. A body, suspended in a two-phase flow, is subject to a variable force, because the momentum flux of the flow changes with time. This is an obvious source of vibration and such vibration is particularly serious when the natural frequency of the element corresponds to the frequency of the variation of the momentum flux. Problems of vibration in axial two-phase flow are discussed by Gorman et al (1970), Cedolin et al (1970), Cedolin et al (1971) and Pettigrew and Paidonssis (1975). Very little information exists on tube vibration in two-phase cross flow; limited experiments by Pettigrew and Gorman (1977) show, however, that the onset of fluid-elastic vibration in two phase cross flow is consistent with values calculated from equation 5.1 using the homogeneous mean velocity and density. The presence of vibration can also affect the local two-phase heat-transfer phenomena (see, for instance, Fuls and Geiger (1970)).

A good general source on vibration problems is the proceedings of the International Symposium on Vibration Problems in Industry, held at Keswick, England, in April, 1973 and May 1978 and organised jointly by the UKAEA and the UK National Physical Laboratory.

Vibration measurement techniques have been widely developed and a review of vibration instrumentation for nuclear reactors is given by Thomas (1973). The instrumentation systems range from optical (interferometry) methods (Pastorius and Pryor (1974)) to various forms of accelerometer probes (Bouche (1967), Cedolin et al (1970) (1971), Epstein (1972), McCalvey and Thompson (1975) and Taylor and Millican (1974), for example) and strain gauges (Pettigrew and Gorman (1977)).

Measurement of tube momentum flux, even in the steady state, is of interest in understanding two-phase flows, since the accelerational component of pressure gradient is given by the rate of change along the channel of the momentum flux M defined by:

$$M = \int_A (G_L u_L + G_G u_G) dA \qquad (5.2)$$

where G_L and G_G are the local mass fluxes and u_L and u_G the local velocities of the liquid and gas phases respectively. Note that for the homogeneous model:

$$M = \frac{AG^2}{\rho_H} \tag{5.3}$$

and for the separated flow model:

$$M = AG^2 \left[\frac{x^2}{\alpha \rho_G} + \frac{(1-x)^2}{\rho_L (1-x)} \right] \tag{5.4}$$

In an annular flow, in which a fraction F_E of the liquid flow is entrained, and in which the liquid film occupies a fraction β of the cross-section, the momentum flux is given by:

$$M = AG^2 \left[\frac{x^2}{\alpha \rho_G} + \frac{(1-F_E)^2 (1-x)^2}{\beta \rho_L} + \frac{F_E^2 (1-x)^2}{(1-\beta-\alpha) \rho_L} \right] \tag{5.5}$$

If the droplets in the core are travelling at the same velocity as the gas then:

$$\beta = 1 - \alpha - \frac{\alpha F_E (1-x) \rho_L}{x \rho_G} \tag{5.6}$$

Measurement of momentum flux can be achieved by converting all of the forward momentum into a force. This is achieved by turning the flow through a right-angle by impingement onto a plate, the force on the plate being equal to the momentum flux. Measurements of momentum flux have been made, for instance, at MIT by Andeen and Griffith (1968) and Yih and Griffith (1970). The apparatus used is shown in Figure 5.1.

The flow is diverted through a right-angle, impinging on a plate mounted on a deflecting beam, the beam deflection being determined by LVDT coils as shown.

Typical results for momentum flux obtained by Andeen and Griffith are shown in Figure 5.2 in the form of momentum multiplier (M/G^2A) as a function of quality. The upper line represents the homogeneous flow model and the lower line a "minimum momentum flux" model. The data agreed better with the homogeneous model (equation 5.3) though Andeen and Griffith are at pains to stress that this does not actually mean that the flow is homogeneous.

88 Measurement of Two Phase Flow Parameters

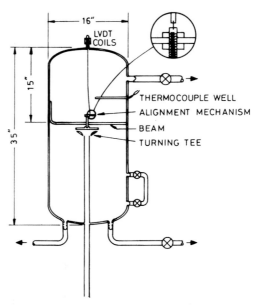

Figure 5.1: Apparatus used by Griffith and co-workers for this measurement of two phase momentum flow.

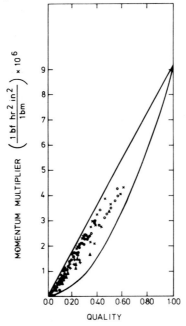

Figure 5.2: Data for momentum multiplier for air-water vertical flow. Andeen and Griffith (1968).

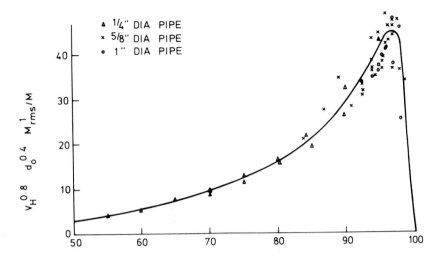

Figure 5.3: Data for r.m.s. fluctuations of momentum flux. Yih and Griffith (1970).

Data on momentum fluctuations are given by Yih and Griffith (1970) and are exemplified in Figure 5.3.

Relatively few investigations have been done of vibrations and the related transient momentum-flux in the two-phase flow, though there is interest in the measurement of momentum-flux from the point of view of transient-flow metering. Further studies should be performed on transient momentum-flux in two-phase flows. Techniques for measurement of momentum flux distribution are reviewed below.

Flow Distribution

In many practical situations, it is necessary to divide a two-phase flow into separate streams (eg the return header of a two-pass air-cooled condenser). Also, in many accident analyses for nuclear reactors, a situation will occur where a two-phase flow divides between a number of reactor channels, during the blow-down phase. It definitely cannot be assumed that the flow distribution is uniform. At each junction in the distributor header, the quality of flow in the T is different from that in the main header. Figure 5.4 illustrates results obtained at Harwell on this separation effect. Similar effects are also reported by Fouda and Rhodes (1972) (1974). For single junction tests, the flows from the two outlets may be separated and the gas and liquid flows measured independently. For multiple junction systems, this is not possible since, in a manifold, the overall distribution results from

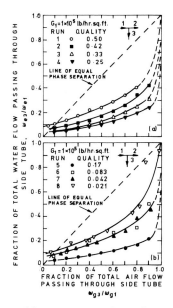

Figure 5.4: Separation effects at a T-junction two-phase flow. (Results from AERE Harwell).

the cumulative contributions of each junction and any flow disturbance would itself change the distribution. The system used at Harwell for measurements in an array of four tubes is illustrated in Figure 5.5.

Flow conditions were set up and the pressure drop (Δp) measured in each tube. The void fraction (α) was determined for each tube by quick closing valves, operated simultaneously by the connected lever system as shown. From the pre-determined relationships between liquid and gas flowrate (W_L and W_G) and α and Δp respectively, cross-plots were prepared which allowed W_L and W_G to be estimated from the measured values of α and Δp. Table 5.1 shows some results obtained for this method. W_G and W_L are the actual flows of gas and liquid, respectively, in each of the tubes and $\bar{W}_G$ and $\bar{W}_L$ are the mean flows per tube. A gross mal-distribution occurs and this will have a profound effect on the design.

Relatively little is known about two-phase flow distribution in manifolds and header systems. The limited information which does exist suggests that gross mal-distributions can occur. Further work needs to be carried out, both on the separation effects at single junctions and on manifold flow distributions in two-phase flow. The effects of interactions between successive junctions are likely to be particularly important when the distance between the junctions becomes small. Such work is relevant to

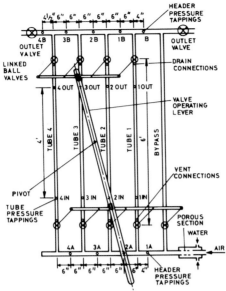

Figure 5.5: Apparatus for measurement of two phase flow distribution in an array of four horizontal tubes connected to common inlet and outlet headers.

TABLE 5.1

Distribution of Two-Phase Flow between Four Parallel Tubes

	Tube Number				Ratio of maximum to minimum flow
	1	2	3	4	
$W_G/\bar{W}_G$	2.05	1.39	0.37	0.19	10.8
$W_L/\bar{W}_L$	0.46	0.46	1.15	1.93	4.2

nuclear reactor problems, due to the possible occurrence of two-phase flow in distribution headers during accident conditions.

Stability

Two-phase boiling and condensing systems are liable to exhibit gross variations in flow, due to feedback, and other forms of instability. Instability mechanisms are described in Chapter 1 and are reviewed by Gouse (1964); a classification of instability types is given by Boure et al (1973). One must ensure that the response and dynamic range of the detection system is sufficient to follow the time variations. An example of such a fast-response system is given by Possa and Van Erp (1965), who used a transient dilatometer method for void-fraction determinations, in an unstable loop. The impedance void-gauge method (see above), is also particularly useful for stability studies and has been used in this context by Dijkman (1970) and Akesson (1968). More recently, large strength gamma sources are becoming available which allow response times down to 1 ms. Examples of recent work on stability are the studies of Aritoni et al (1977), Kakac et al (1978), Rakopoulas et al (1978) and Fukuda and Kobori (1978).

Further studies of two-phase instability are required in those areas where the assumptions of current models (homogeneous flow) are likely to be most significantly in error.

Quality and Mass-Flow Measurement

The response of nearly all metering devices, for a two-phase flow, will depend on the mass-flow and the quality. In principle, therefore, if the mass flow is known, then the quality can be determined from the response of the metering system, and vice versa. If neither mass-velocity nor quality are known, then two separate measurements normally need to be made, in order to deduce one (or both) parameters. Measurements of transient mass flow in nuclear blowdown experiments have assumed great importance in recent years. These normally rely on two simultaneous measurements and, as we shall see below, this introduces considerable uncertainties since accurate in situ calibrations *for flow conditions corresponding to those found in the transient* is not possible. This has led to the search for metering systems which measure the mass flow specifically - ie, the so-called "true mass flow meters". These are discussed further below, but have not yet reached a point of development where they are widely applied. A review of mass flow measurement in two phase flow is given by Banerjee (1978).

Since, in general, two measurements are required to determine mass flow and/or quality when both parameters are unknown, the general practice has been to use void fraction measurements (normally by means of gamma ray or X-ray absorption - see Chapter 3) coupled with some flow dependant measurement. The choices made for the second measurement have included:
(1) Differential pressure devices (orifices, venturies etc). The difficulty here is that the device itself provides a flow resistance which may change the experimental result if blowdown is being studied. However, differential pressure devices are suitable for steady state measurements.
(2) Turbine flow meters. Here, the flow drives a turbine

and the rate of revolution gives an indication of the
flow rate.
(3) Measurement of the thrust at discharge in blowdown.
This technique has been used by, for example, Hutchson
et al (1975).
(4) Tracer techniques. Here, a tracer is introduced into
the fluid and its motion followed by various methods-
(5) Drag screens. Here, the force is measured on a screen
placed normal to the flow.
(6) Ultrasonic flowmeters. The application of these devices
to two phase flows is described, for instance, by
Lynnworth et al (1976). These instruments are likely
to be very sensitive to the specific flow pattern.

Of the above, the differential pressure, turbine flow meter, tracer and drug screen methods have been applied most widely and are discussed in more detail below.

Ideally, the second measurement device should measure some property of the full system flow. However, in large scale nuclear blowdown experiments such as the LOFT experiment in the USA (McPherson (1977)), the pipe diameters are so large that the use of devices such as full flow drag screens is scarcely feasible and there would, in any case, be calibration problems (see below). Here, it is the usual practice to employ instrumentation "rakes" incorporating several localised devices. Since devices with good transient response are required, the choice for these experiments has been local drag disks or screens mounted together with a small turbine meter as illustrated in Figure 5.6. Three such units are mounted across a diameter of the outlet pipe and a three-beam gamma densitometer is also used, with the beams passing through the axial loci of the drag disk/turbine flowmeter units. The development of this measuring system is described in detail by Wesley (1977) and by Fincke et al (1978). A further discussion of local drag discs and of multibeam gamma densitometers is given in Chapter 6. The reader is also referred to the section of Chapter 6 which discusses the question of locaised mass flow measurement; clearly, total mass flow can be deduced by integration of local mass flow measurements of any kind, but many techniques for local mass flow measurement are not readily adopted to transient studies.

Several important general points need to be made about mass flow measurement using two devices together:
(1) The response of metering devices in two phase flow
tends to be highly sensitive to the local flow pattern.
This in turn is sensitive not only to the mass flows
of the gas and liquid phases but also to the upstream
configuration and flow history. An illustration of
this point will be given below in the context of flow
measurements using a venturi.
(2) The best practice is to calibrate the instruments
with known phase flow rates and *with an exact simula-*

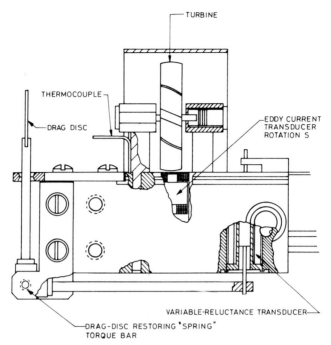

Figure 5.6: Drag disc/turbine flow meter combination used in LOFT tests (Wesley (1977)).

tion of the upstream pipework. However, the question always has to be asked: How far back upstream is it necessary to simulate? There is no general answer to this question and it can only be partly resolved by trial and error.

(3) In the case of transient tests, the flow pattern is likely to be time dependent in addition to being upstream history dependant. Furthermore, it is quite likely that there are departures from thermodynamic equilibrium and the instruments themselves may promote local interphase heat and mass transfer. The uncertainties in transient tests are yet another reason for going to more localised measurements as indicated above. The argument is that the summation of results from localised instruments is more reliable since the variation of flow pattern *across the instrument itself* is likely to be less severe.

(4) Though the ideal is to use in-situ independant calibration, the more usual method is to interpret the measurements from an instrument in terms of a theory whose validity is tested by conducting separate experiments. Very often, these experiments are for very different flow conditions and may even be for different fluids; for example, instruments used for steam/water

flows are often tested using air water flows.
(5) The responses of instruments may be different in transient situations. It may sometimes be possible to correct for this and an example here is the turbine flowmeter, as discussed below.

Bearing all these factors in mind, it is scarcely surprising that equivocal results are often obtained. Since mass velocity can be determined from any pair of measurements, tests in which several combinations are possible are particularly valuable in assessing the general validity of the results. Figure 5.7 illustrates data obtained for the LOFT tests in which, in addition to the "rake" described above (which offers the pairs: drag disk/densitometer, turbine meter/densitometer and drag disk/turbine), an overall differential pressure over a section of the loop was measured which, coupled with a density measurement, could also give the flow rate. The large deviation between the various pairs of instruments themselves and with the known initial total mass in the system, will be noted. Much further development is needed if satisfactory mass flow measurements are to be achieved under these conditions.

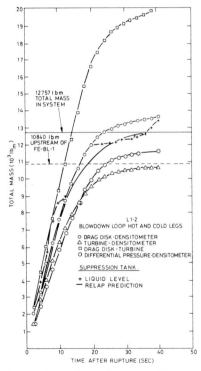

Figure 5.7: Comparison of predicted mass efflux in a LOFT test by using several combinations of pairs of instruments (Beardon (1977)).

The measurement of the quality of high quality (wet steam) flows is discussed by Ryley (1966), Claverie (1970), Rutz (1971) and Ryley and Holmes (1973). The methods used for wet steam are usually based on sampling techniques, though, in order to correct for the error due to the flow of part of the liquid on the tube wall, it is often necessary to have a film stripper in addition to sampling probes. A further discussion of mass-flow sampling and liquid film-flow measurements, respectively, is given in Chapter 6.

Differential Pressure Devices

The simplest form of metering device for quality and mass flow (when used in combination with another measurement) is a differential-pressure instrument. The use of orifice plates for quality metering is described by James (1965-66), Bizon (1965), Heckle (1970), Collins and Gacesa (1970), Rooney (1972), Dickson and Wood (1972) and Chisholm (1972) use the following expression for calculating the orifice pressure drop Δp_{TP}:

$$\frac{\Delta p_{TP}}{\Delta p_L} = 1 + \frac{C}{X} + \frac{1}{X^2} \qquad (5.7)$$

where Δp_L is the pressure drop for the liquid phase flowing along, X is the Martinelli parameter (equal to $(\Delta p_L/\Delta p_G)^{\frac{1}{2}}$ where Δp_G is the pressure drop for the gas phase flowing alone). The parameter C is given by Chisholm as:

$$C = \frac{1}{S}\left(\frac{\rho_L}{\rho_G}\right)^{\frac{1}{2}} + S\left(\frac{\rho_G}{\rho_L}\right)^{\frac{1}{2}} \qquad (5.8)$$

where S is the "slip ratio" - ratio of gas and liquid velocities - and ρ_L and ρ_G the liquid and gas densities. For homogeneous flow, S = 1 and such flow is often assumed in interpreting orifice data. However, this assumption can be considerably in error as shown by Rooney (1973). Pressure drop fluctuations in two phase flow through orifices are discussed by Ishigai et al (1965). The use of venturi meters is discussed by Smith et al (1962), Shires (1966), Harris (1967), Bizon (1965), Palm et al (1968), Collins and Gacesa (1970), Alekseev et al (1973), Harris and Shires (1972), Fouda and Rhodes (1976) and Sheppard et al (1977). Harris and Shires use an analysis of the Chisholm type as described above whereas Sheppard et al use a separated flow model giving the pressure drop Δp_{TP} (convected for static head changes) for the venturi as:

$$\Delta p_{TP} = C_1 \alpha u_L^2 [\rho_G S^2 + \rho_L (1-\alpha)/\alpha] \tag{5.9}$$

where C_1 is a constant for the venturi and u_L the liquid phase velocity.

Colombo and Premoli (1972) suggest the use of differential-pressure measurement across simple pipe-elbows as a means of measuring quality.

Provided that the differential-pressure instruments are properly calibrated, a good degree of accuracy can be obtained in the measurements. However, care should be taken with this, and all other forms of flow-and quality-metering, to ensure that the upstream flow conditions are reproducible between the calibration and the actual application. Harris (1967) shows that the calibration of a venturi-meter, for steam-water mixtures, is affected by the geometry of the upstream pipe-work. Results which illustrate these effects are shown in Figure 5.8; for steam-water flow at a given pressure, the data for differential pressure could be correlated in terms of the product $W^2 x$ where W is the mass rate of flow and x the quality. However, the relationship obtained depended on the upstream pipe configuration as illustrated.

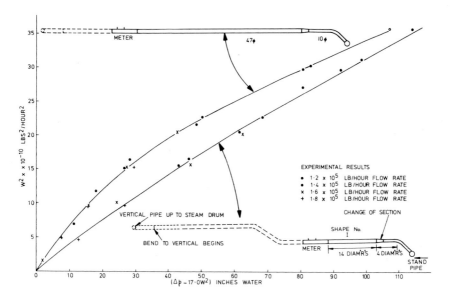

Figure 5.8: Effect of upstream pipe configuration on calibration of a two phase venturi for steam/water flow at 70 bar pressure. (Harris (1967)).

Turbine Flow Meters

Another form of metering system used in flow/quality metering is the turbine flow-meter. This has the advantage of fairly rapid response (ie about 30 ms in liquids, but probably longer in two-phase flow) but, again, care has to be taken in the interpretation of the data for two-phase flows. Reviews on the use of turbine flowmeters, in single-phase flow, are given by Strohmeier (1974) and Galley (1974). The development of miniaturised meters for sub channel measurements is described by Piper (1974b). Results obtained using turbine flow-meters for two-phase flow, are reported for instance by Clark (1946), Flinta et al (1971), Rouhani (1964) (1974), Arave and Goodrich (1974), Wesley (1977), Beardon (1977), Banerjee et al (1978) and Banerjee (1978). Rouhani presents a formula for the calculation of the rotational speed C of a free turbine flow-meter in two-phase as follows:

$$C = kWv_{TP} \qquad (5.10)$$

where k is a proportionality constant, W is the mass flow and v_{TP} is a two-phase "momentum" specific volume calculated from:

$$v_{TP} = \frac{x^2}{\alpha \rho_G} + \frac{(1-x)^2}{(1-\alpha)\rho_L} \qquad (5.11)$$

Where x is the quality, α the void fraction and ρ_G and ρ_L are the gas and liquid phase densities. α is a further unknown and must be estimated or measured. Alternative models for turbine flow meter response are discussed by Banerjee (1978). A specific difficulty in using turbine flow meters in transient conditions is that, with a change in flow conditions, there is a change in the angular momentum stored in rotor and in the associated fluid rotating with the rotor. The speed of rotation, therefore, does not represent the instantaneious value of the mass flux. Kamath and Lahey (1977) present an analysis of the transient response of a turbine flow meter; the solution is non-linear and fairly complex but a computer code was developed which could predict the true temporal variation. Figure 5.9 shows some results calculated by Kamath and Lahey (1977) using this code for a typical blowdown transient. As will be seen, the difference between the "quasi-steady state" interpretation and the true mass flux depends on the moment of inertia (I) of the rotor. Note how the mass flux is first underpredicted and then over-predicted as the rotor angular momentum increases and then falls during the transient. Lassahn and Arave (1974) experienced a high percentage of failures of their turbine

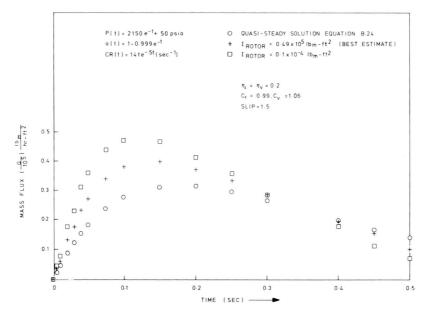

Figure 5.9: Response of turbine flow meter to blowdown transient. Calculations of Kamath and Lahey (1977).

flowmeter. In the same tests, on LOFT fuel assembly models, a thermocouple failed due to its sheath splitting. The cause of the latter is not explained, but foreign matter, in the flowing medium, caused the jamming and excessive wear of the turbines. Turbine flow-meters, sampling the whole flow, are likely to perform impredictably when they are used for two-phase flow regimes other than those approximating to homogeneous. For instance, in annular flow, the response is likely to be different to homogeneous flow at the same void fractions and mass flux.

Tracer Techniques

Another measurement system, of potential use in flow and quality metering, is that of tracer injection. This is widely used in the metering of single-phase flows (see, for instance, Clayton (1960) and Uhl (1974)), and has been applied to two-phase flow systems by James (1961), Forslund and Rohsenow (1966), McLeod et al (1971), Stothart (1972) and Brown (1974a), (1974b). Using this technique, it is possible to separately label the liquid and vapour phases (Brown (1974a) (1974b)). The potential of the technique appears to be large but there are difficulties in its application due to:
(1) The problem of obtaining a uniform distribution of the tracer throughout the flowing fluid.
(2) Interphase mass transfer and mixing giving rise to a complex modulation of the input tracer pulse. This

aspect is thus discussed further in Chapter 6.
(3) Changes in flow pattern and velocity between the
point of tracer addition and the point of detection.

It is difficult to obtain uniform tagging with conventional techniques and the complex response invalidates the use of interpretive methods developed for single phase flow studies. However, recently, there has been a new development which overcomes some of the major difficulties in tracer methods. This new method involves the use of pulsed neutron activation (PNA). A pulse of 14MeV neutrons is generated from a special neutron source (Kehler (1978)) or from an accelerator (Lahey (1978)) and irradiates a short section of the tube carrying the flow. The neutrons produce a short-lived radioisotope of nitrogen by the reaction $O^{16}(n,p)N^{16}$. The half-life of N^{16} is 7.24 seconds and, in its decay, it produces gamma rays of energy 6.1 MeV. A gamma ray detector is placed at a distance Z_o downstream of the irradiation section and the count rate, $c(t)$, determined as a function of time. Specifically, the total counts received by the detector is measured in a series of time intervals Δt, the counts received in the ith interval being given by:

$$C_i = \int_{(i-1)\Delta t}^{i \Delta t} c(t) dt \qquad (5.12)$$

where $t = 0$ represents the time of the activation pulse. Kehler (1978) implicitly relates the mass flux, G, to the counts C_i by the relationship:

$$G = K \sum_{i=1}^{\infty} \frac{1}{t_i} C_i e^{\lambda t_i} \qquad (5.13)$$

where t_i is the time at the mid-point of the ith interval, λ the decay constant of the activity and K a constant for the system which can be determined by calibration with a single phase water flow. By estimating the total counts generated during the activation process, it is also possible to estimate the mean density in the channel (Kehler (1978)). The PNA technique is a major advance in tracer methods and shows promise as a standard method for two phase mass flow.

Full Flow Drag Screen

A drag screen (typically a perforated plate) may be mounted in such a way that it occupies the full cross section of the pipe. The screen is attached to a beam whose deflection, due to the fluid forces on the screen, may be measured. Figure 5.10 shows a typical arrangement.

Ideally, each part of the device operates with a constant

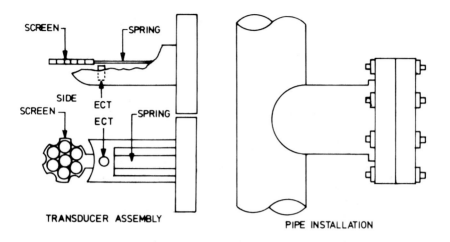

Figure 5.10: Drag screen mounting arrangement (Arave et al (1977)).

drag coefficient so that the force on the part is proportional to G^2/ρ, where G and ρ are the local mass fluxes and densities. Thus, the force on the whole screen can be equated with $\bar{G}^2/\bar{\rho}$ where $\bar{G}$ and $\bar{\rho}$ are average values. The average density $\bar{\rho}$ can be measured using, say, gamma densitometry and this allows the estimation of $\bar{G}^2$. Good results using this instrument are reported by Arave et al (1977) but one could expect difficulties in the averaging process if the flow were markedly non-uniform.

True Mass Flow Meters

The importance of, and difficulties in, mass flow measurement has led to the search for measurement principles which are dependent only on the mass flow. Two devices of this type which show some promise are:
(a) *Gyroscopic/coriolis mass flow meters*. This device uses a C-shaped pipe as shown in Figure 5.11. The pipe is oscillated, using the magnetic force and spring device, in the direction normal to the plane of the C-pipe. This induces a Coriolis force on the particles flowing through the tube and this causes torsional oscillation of the tube which can be measured. The moment of the Coriolis force is proportional to the mass flux through the tube. A detailed description of a device of this type is given by Smith (1978).
(b) *True turbine mass flow meter*. This device is described by Barshdorff et al (1978) and is illustrated in

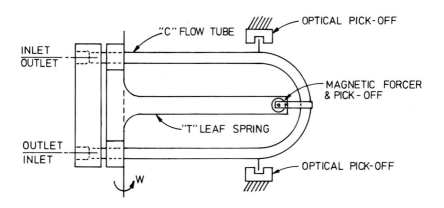

Figure 5.11: Gyroscopic/Coriolis mass flow meter (Smith (1978)).

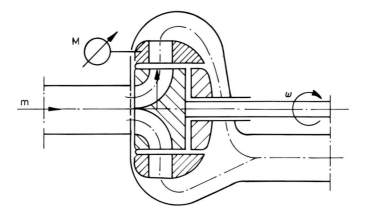

Figure 5.12: True turbine mass flow meter (Barschdorff et al (1978)).

Figure 5.12. The two-phase fluid is passed into a multiblade turbine with a rotor of diameter r rotating at ω radians/sec. Fluid leaving the rotor impinges on the stator and imparts its angular momentum to it generating a torque M on the stator, which is measured. Ideally, M is given by $Wr^2\omega$ where W is the mass rate of flow. In practice, the blade efficiencies of the rotor and stator are less than unity and, typically, the torque is 6% lower than the value given by $Wr^2\omega$.

The second of these devices has recently been used successfully in blowdown experiments giving good agreement with known mass inventories when integrated over the total blowdown transient.

To summarise, one can say that, properly calibrated, instruments of the orifice/venturi type can give a good accuracy. However, care should be taken to ensure that the upstream flow conditions are reproducible between the calibration and actual application. Turbine flow-meters are likely to be extremely sensitive to flow patterns and the use of tracer techniques may be limited due to difficulties in interpretation. There is a growing need for an instrument for measuring mass flow/quality in two-phase systems, which is insensitive to inlet flow configuration and there are some promising new developments which may partially fulfill this need.

Liquid Level Detection

During transients, there is often a requirement to establish, accurately, the changes in liquid level within vessels. In the LOFT reactor experiment, the water level in the down-comer annulus and the lower plenum will be measured during LOCA transients. Arave (1972b) describes an impedance device for level determination, which consists of a cylindrical electrode surrounded by a grounded cylinder. Water entering the annular space changes the conductivity. Aerojet Nuclear Company (1974) describe an alternative impedance technique. The detector consists of a 1/8 inch diameter electrode, surrounded by a cylindrical set of grounded electrodes (made up from the cable sheaths). A level probe can be made up of between 2 and 21 electrodes, each spaced a minimum of 3.5 inches apart, axially. The presence of water or steam at any location along the probe is determined by measuring the conductivity of the medium surrounding the respective electrodes. The impedances are measured using a 200 Hz current. In flashing water, the impedance is intermediate between that of water and that of steam. There is a time lag of a few seconds for the ceramic seal to dry off, before an infinite impedance is recorded for an all-steam atmosphere.

Lindstrom et al (1970) describe a technique which makes use of the fact that an electric pulse, travelling along a transmission line, is reflected from any point on the line, where the line impedance changes suddenly, eg where a fluid is present in the line. High accuracy was claimed but the performance of such a device in the highly disturbed interfaces expected in blowdown, may not be so satisfactory.

Larsen (1975) mentions the use of a conductivity-sensitive probe, as a liquid-level detector, but does not describe the device in other than general terms. He refers to it as very durable and able to withstand repeated thermal shocks.

Lehmkuhl (1972) used a pulse-reflection technique - termed, "time domain reflectometry" - as a tool for the measurement of liquid level. With pure, or nearly-pure, water, his instrument gave a good and reproducible match between readings from pulse-measurement and direct sighting (within 3 mm). However, the presence of electrolyte in the reflectomer cell gave errors of several centimetres, due to the use of water as a calibration fluid.

All devices for liquid level are subject to the difficulty that the water/steam interface is unlikely to be well defined in practice, particularly during a blow-down. Care should be taken to ensure that the conditions around the level detector are representative of those in the bulk of the vessel. Further measurements are desirable in which various liquid level devices are compared under transient conditions.

6 Second Order Parameters

Introduction

   A 'second order parameter' is a time or space averaged parameter which is of interest in research on the system aimed at improving understanding (usually with the objective of improving methods for predicting the first order - design - parameters). The items considered under this heading in this chapter are as follows:
(a) Flow pattern
(b) Time averaged film thickness and amplitude distribution.
(c) Mass flow distribution
(d) Phase concentration distribution; local void fraction.
(e) Velocity and momentum flux distribution.
(f) Concentration distribution.
(g) Mixing characteristics of two phase flow including phase mass transfer.
(h) Drop, bubble and particle size measurements.
(i) Wall shear-stress.
(j) Temperature distribution.
(k) Entrainment - film flow rate.
(l) Contact angle.

Flow Pattern

   As described in Chapter 1 it is often helpful in analysing two phase flow systems, to classify the flow into a number of "flow patterns" or "flow regimes". This helps in obtaining a qualitative understanding of the flow and also, hopefully, ultimately should lead to better prediction methods for the various two-phase flow parameters. Typical classifications of two-phase flow patterns are discussed by, for instance, Hewitt and Hall-Taylor (1970).
   The most straightforward way of determining the flow pattern is to observe the flow in a transparent channel, or through a transparent window in the channel wall. However, the phenomena often occur at too high a speed for clear observation and high-speed photography or related techniques must be used. Developments in the photography of two-phase flow are described by Cooper et al (1963) and Arnold and Hewitt (1967). Examples of flow-pattern studies using photographic techniques for gas-liquid flows,

are those of Hewitt and Roberts (1969), Bergles and Suo (1966), Raisson (1965), Hsu and Graham (1963), Traviss and Rohsenow (1973), Mayinger et al (1976), Schlapbach (1977) and Kordyban (1977). Photographic studies of flow patterns in liquid-liquid flows have been reported by Hassan et al (1970). The problem of observing patterns in heated tubes can be overcome by using transparent, electrically-heated walls (Gouse and Dickson (1966), for instance).

Unfortunately, even with high-speed photography, it is often not possible to observe the structure of the flow clearly, due to the complex light-refraction paths within the medium. In these cases, X-ray photography can be very useful and developments of this technique are described by Derbyshire et al (1969), Hewitt and Roberts (1969) and Mayinger and Zetsmann (1977). The flow-pattern map, shown in Figure 6.1 was obtained on the basis of X-ray photographs, coupled with high-speed flash and cine-photography. X-ray fluoroscopy can be used directly for visual observation of two-phase flows, though the time resolution is obviously not as good as flash X-radiography. Fluoroscopy methods for the study of Freon evaporation, in a tube, are described by Johanns (1964) and Baker (1965).

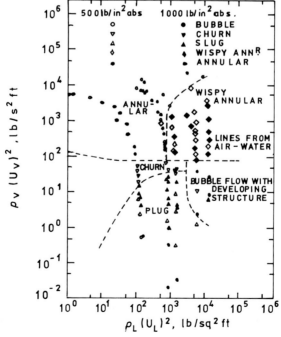

Figure 6.1: Flow pattern diagram based on X-ray and visual studies (Hewitt and Roberts (1969)).

The general unreliability of photographic methods has led some authors to seek other techniques for flow pattern categorisation. The most popular of these is to insert a needle facing directly into the flow and to measure the current from the tip of this needle, through the two-phase flow, to the wall of the channel. The current is displayed on an oscilloscope and the type of response is considered to be representative of the flow pattern. For instance, if no contacts are made between the needle and the wall, one may assume a continuous gas core and, thus, annular flow. High frequency interruptions of the current indicate bubble flow, and so on. This technique has been used by a number of authors including Katarzhis et al (1955), Solomon (1962), Griffith (1963), Wallis et al (1963) Haberstroh and Griffith (1965), Raisson (1965), Bergles and Suo (1966), Ferrel and Bylund (1966), Fiori and Bergles (1966), Bergles et al (1967) (1968), Bergles (1967), Beckerleg and Kiel (1970), Janssen (1970), Janssen et al (1971), Pinczewski and Fell (1972), Dsarasov et al (1974), John et al (1976), Kikuchi et al (1975) and Reiman and John (1978). Detailed comparison between results obtained with this technique and those obtained by visual methods, are reported by Bergles and Suo (1966) and Raisson (1965). Although the visual and contact methods agree moderately well, where the flow regime is clearly defined, discrepancies arise in the transition regions, particularly in the transition from semi-annular to annular flow. If one accepts a definition of annular flow that specifies a continuous gas core, the needle-contact probe, even in the centre of the channel, would still give a response due to impingements with the tips of protrusions or waves, and might indicate no annular flow when it is, in fact, occurring. The X-ray photography method is, therefore, more reliable in examining these regions. Ryan and Vermeulen (1968) describe a multiple-wire conductance-probe device for detecting slug flow.

Four other techniques can be mentioned here, which show some promise in regime delineation:
(1) Electro-chemical measurement of wall shear-stress. Kutateladse (1973) shows that the wall shear-stress fluctuations are related to the flow pattern, in heated two-phase flow.
(2) X-ray fluctuations. This method for flow pattern delineation is described by Jones and Zuber (1974) and is illustrated in Figure 6.2. The method uses the instantaneous measurement of void fraction, using X-ray absorption, as a means of defining the flow pattern. The probability density of the instantaneous void-fraction is obtained from the output and shows characteristic peaks in the low and high void-fraction regions for bubble and annular flow, and shows a double peak for slug flow. This method is very promising but has yet to be applied to conditions of

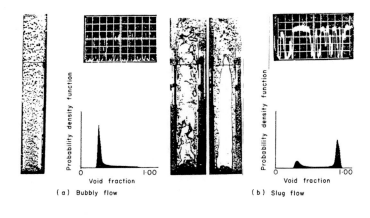

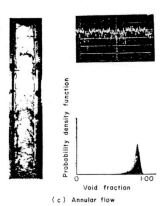

Figure 6.2: Use of fluctuations in X-ray absorption to delineate flow patterns (Jones and Zuber (1974)).

high mass velocity, where all the major problems of delineation occur.
(3) Analysis of pressure fluctuations. Measurements of fluctuating pressure have been used to identify flow regimes by Hubbard and Dukler (1966) and by Simpson et al (1977). Hubbard and Dukler distinguished essentially three types of power spectral density distribution obtained from pressure fluctuation measurements in horizontal air-water flows. These are illustrated in the sketch in Figure 6.3 and are as follows:
  (a) *Separated flows*. Here, there is a peak at zero frequency; this type of response is obtained from stratified, wavy and cresting flows.
  (b) *Dispersed flows*. Here, there is a flat, relatively uniform, spectrum.
  (c) *Intermittent flows*. Here, there is a characteristic peak in the power spectral density and this kind

Flow Pattern

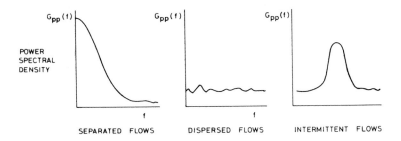

Figure 6.3: Relation of power spectral density function and flow type (Hubbard and Dukler (1966)).

of result is obtained for plug, slug and slug/annular flows.

Care has to be exercised in the interpretation of pressure fluctuations since reflections from the outlet can be important. This point is discussed further in Chapter 7.

(4) In horizontal tubes, the flow is assymetric and the distribution of void fraction can give important clues about the flow pattern. A multi-beam X-ray method has been devised for this purpose and is described by Smith (1975). The system is illustrated in Figure 6.4. It was used for evaluation of flow patterns in blowdown from a horizontal tube. Differences between the mean void fraction, recorded along the various beams, were used in deducing the developing flow pattern. A similar method can be applied using gamma absorption and, with the recent development of high-strength gamma sources, this method can be used for rapidly-developing flows such as those found in blowdown. Multi-beam gamma absorption devices are described, for instance, by Ybarrondo (1975), Lassahn (1975a), Lassahn (1977), Reiman and John (1978) and Wesley (1977). Perhaps the most fully developed multi-beam system is that used for the LOFT blowdown tests in the USA. A diagram of this unit is given in Figure 6.5 and an analysis of the response of this system (from which density distribution information can be obtained) is presented by Lassahn (1977). A 30 curie Cs-137 source is used giving good time response and the application of the densitometer to flow regime detector is illustrated in Figure 6.6. The use of this system, coupled with drag discs and turbines, in estimating mass flow was discussed above. Although these devices are helpful in providing extra information, it should be realised that ambiguities still exist and two phase flow

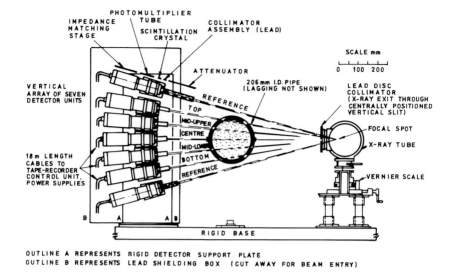

Figure 6.4: Multi-beam X-ray system for determination of chordal mean void fractions (and, hence, for determination of flow patterns) during blowdown from a horizontal tube (Smith (1975)).

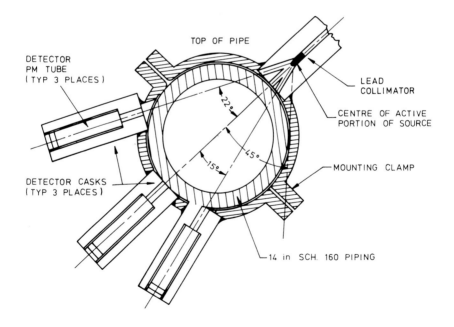

Figure 6.5: Three beam gamma densitometer used in LOFT tests (Wesley (1977)).

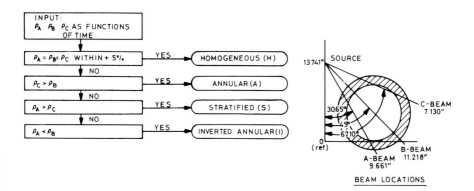

Figure 6.6: "Truth diagram" for determining flow regime for the response of the LOFT three beam gamma densitometer.

patterns can actually be very complex.

The delineation of flow patterns is important in the development of better models for two-phase flows but data on flow patterns should be treated with reservation. It would seem that the best results can be obtained using visual techniques, providing they are coupled with X-ray observations.

The general state of knowledge on flow patterns is unsatisfactory and an improvement is needed in the development of better models for two-phase flows. Existing data on flow patterns should be treated with reserve due to the lack of specificity of the various techniques available. Of the more "quantitative" techniques available, the ones showing most promise appear to be those involving the analysis of X-ray absorption fluctuations and interpretations of multi-beam absorption.

Further development should be encouraged of the X-ray fluctuation technique for flow pattners. Also, a much more detailed study is needed of flow patterns in the high mass-flow region, possibly using flash X-radiography and multi-beam absorption.

Time-Averaged Film Thickness and Amplitude Distribution

In many two-phase gas-liquid and liquid-liquid systems, a liquid film exists on the channel wall which can play an important role in the mass-momentum- and heat-transfer processes in such flows. A knowledge of the time-averaged value of the thickness of this film, and its probability distribution, is useful in testing theoretical models. Comprehensive reviews of measurement methods for liquid film-thickness are given by Collier and Hewitt (1964) and Hewitt and Hall-Taylor (1970). Collier and Hewitt classified the techniques used in film thickness measurement as follows:

(1) *Film average methods*. These include measurement of holdup (with quick closing valves), techniques for weighing the film on the surface, and whole-film conductance measurement.

(2) *Localised methods*. These include the use of localised film conductance measurement, film capacitance measurement, broad beam radioactive absorption methods and measurement of the emission from a radioisotope dissolved in the film liquid.

(3) *Point measurements*. These include the popular needle contact method, the light absorption method, the fluorescence method and the narrow beam X-ray absorption method.

This classification is still valid for the more recent work though there is little interest nowadays in film average methods. Of the localised methods, the film conductance technique has proved by far the most popular and, of the point methods, the most popular method, for mean thickness and probability distribution of film thickness, has been the needle contact technique. Both these technqiues are discussed in more detail below. The film conductance method can be used to give a continuous record of film thcikness, which is useful in studying interfacial wave behaviour. However, for average and instantaneous measurement, the fluorescence and X-ray methods offer the advantage of approaching point measurement conditions and will also be described in more detail below; the light absorption method can sometimes be used effectively but it tends to suffer from ambiguity due to interfacial refraction effects and cannot be recommended generally. The radioactive absorption and emission techniques suffer from ambiguities due to counting statistics, self absorption and detector geometry and are not recommended for use where one of the other techniques is feasible.

Film conductance method. This consists of measuring the conductance between two electrodes passed through the (non-conducting) wall of the channel, so as to be flush with the inner surface. The technique has been used by, for instance, Hewitt et al (1962), Collier and Hewitt (1964), Adorni et al (1961b), Gill et al (1963a) (1963b), Puzyrewski and Jasinski (1965), Konobeev et al (1957), Webb (1970b) (1970c), Shiralkar (1970), Telles and Dukler (1970), Butterworth (1968), Coney (1971) (1972), Chu and Dukler (1974), Tomida and Okazaki (1974), Thwaites et al (1976), Chen et al (1977) and Solesio et al (1978). A general review of conductivity probes for two-phase flow is given by Bergles (1969).

A number of probe geometries have been used for the conductance probe method, as illustrated in Figure 6.7. An insurmountable problem, with this type of probe, is that the closer the electrodes are brought together, the smaller is the maximum film thickness that can be measured sensitively. Another difficulty, particularly when investigations are required on a variety of physical properties, is that the liquid must be conducting. However, the probes may be used in a capacitive mode in some applications.

The mean conductance observed gives a measure of the mean film thickness only if the probe is operating in the linear region, whose extent decreases with increasing probe spacing. The linear region may be extended by processing the signal (Webb (1970b) (1970c)) through an analogue function-generator.

The signals from conductance probes can also be used to yield information on the probability distribution of film thickness. Typical data, obtained by Webb, for film-thickness probability distribution are shown in Figure 6.8. The change with distance from the point of injection illustrates the formation of interfacial waves. The mean thickness of the film can be obtained by electronic or graphical averaging. A thorough analysis of the response of conductance probes is given by Coney (1972).

Needle contact method. The needle-contact method consists of bringing up a needle to the film surface and noting the distance from the wall at which the needle makes contact with the film. Since there are surface waves on the film in most cases, the needle may touch the film only intermittently. A distribution of film thickness can

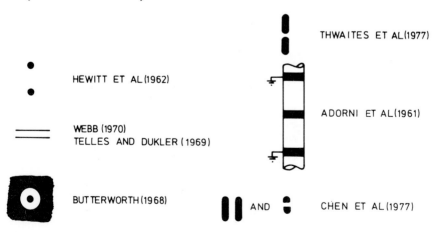

Figure 6.7: Types of probe geometries used for the conductance method for film thickness.

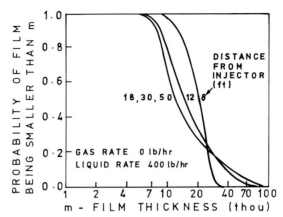

Figure 6.8: Film-thickness probability data obtained by the conductance method (Webb (1970)).

be obtained by measuring the time of contact of the needle with the film, as a function of distance from the wall. This technique is easy to set up, and, although the necessary electronics can be complicated, the system is usually cheap. Experimental work using this technique has been described by Hewitt et al (1962), Schraub et al (1969), Bergles and Roos (1968), Truong Quang Minh and Huyghe (1965), Bergles et al (1966-67), Butterworth (1968), Davis and Cooper (1969), Permyakov and Podsushnyy (1971), Jensen et al (1971), Janssen et al (1971), Kozeki (1973), Dsarasov et al (1974), Maddock and Lacey (1974), Subbotin et al (1975), Kirrilov et al (1978) and Subbotin et al (1978). A typical unit is illustrated in Figure 6.9. The argument against the needle-contact probe has always been that there must exist a possibility of hysteresis in

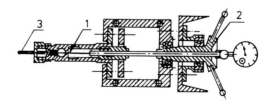

Figure 6.9: Typical needle contact device used for film thickness distribution measurements (Kirillov et al (1978)). (1 - Needle, 2 - Driving mechanism, 3 - wall electrode).

the contact. Some claim that this hysteresis is small, particularly when the gas velocity over the film is high. To the author's knowledge, no positive proof has been obtained one way or the other. However, for the needle-contact probe, there are no limits on the sensitivity. Very thin and very thick films can be measured readily.

Fluorescence technique. Instantaneous and highly localised, measurements of film thickness can be obtained using the fluorescence technique. The principle of the method is illustrated in Figure 6.8. Blue light from a mercury vapour lamp is passed through a microscope illuminator and focussed in a conical beam into the liquid film. The circulating water in the apparatus contains fluorescein dye-stuff and the incident beam excites a green fluorescence in the film. The fluorescent illumination is separated in a spectrometer and its intensity is a direct measure of the liquid film thickness. Detailed measurements using this technique are reported by Hewitt and Nicholls (1969) and by Anderson and Hills (1974). The latter authors used a filter system rather than a spectrometer and their apparatus is described in Chapter 7.

Recently, the theoretical basis for the fluorescence technique has been investigated by Azzopardi (1978) who derives the following expression for the intensity, I,

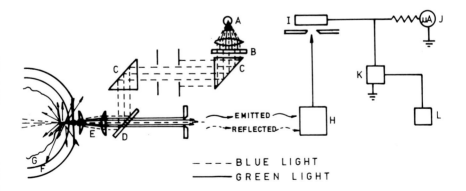

A. MERCURY VAPOUR LAMP
B. FILTERS
C. PRISMS
D. HALF-SILVERED MIRROR
E. OBJECTIVE LENS
F. TUBE WALL
G. LIQUID FILM
H. BARFITT SPECTROMETER
I. PHOTOMULTIPLIER
J. AMPLIFYING MICROAMMETER
K. RECORDING OSCILLOSCOPE
L. MULTICHANNEL SWITCH

Figure 6.10: Fluorescence technique for film thickness measurement (Hewitt et al (1964).

of fluorescent light received by the detector:

$$I = \int_0^m \frac{EaI_0 k_1 cy}{1+k_2 c} \cdot \exp[-k_1 cy - k_3\sqrt{cy} + k_4 cy] dy \qquad (6.1)$$

where I_0 is the incident intensity, E the fluorescence efficiency, a the fraction of fluorescence received by the detector, k_1 the absorption coefficient for the incident light, k_2 a self-quenching coefficient and k_3 and k_4 self-absorption coefficients. y is the distance from the wall, m the film thickness and c the concetration of the dyestuff. Azzopardi solved Equation 6.1 by numerical integration and obtained excellent agreement with the original calibration data.

Further development of the fluorescence technique is proceeding at Harwell with the objective of applying it to non-aqueous fluids and of developing advanced detection methods.

X-ray absorption method. Recent developments of this technique have been made by Solesio et al (1978). By using a thin pencil of X-rays and by using a reference beam technique to overcome the tube, Solesio et al were able to measure 1mm thick films to a precision of better than ±50μm. Comparisons between this method and the film conductance method indicated quite good agreement.

Average film thickness and probability distribution can be obtained most conveniently using film conductance methods, for those cases where it can be applied. For channels with metal surfaces (as used, for instance, in electrically-heated boiling experiments) the needle-contact method may offer advantages. For non-conducting fluids, the fluorescence technique may be the best available, although it is more difficult to apply than the conductance and needle methods. Recent developments of the X-ray absorption technique show that reasonable accuracy can be obtained with films of about 1mm thickness.

Mass-Flow Distribution

The distribution of mass flow in a two-phase gas-liquid, or gas-solid flow, is of considerable interest in interpreting the behaviour of such flows. For large bore pipes, the measurement of overall (including transient) mass flow may also involve the use of distribution measurements which are then integrated to give the required total flow; this point was discussed in Chapter 5.

The local mass-flow of the heavier phase can often be determined using a sampling probe, which consists of a sampling tube facing the oncoming flow and through which a sample of the flow is taken. In practice, the rate of sampling of the heavy, dispersed phase is independent, over a wide range of the rate at which the lighter phase is simultaneously sampled. For example, Gill et al (1963a) (1963b) have shown that the rate of liquid sampling in the gas core of annular flow was independent of the rate of gas sampling for an order of magnitude variation of the latter. A sampling probe is thus often a viable method of measuring the flow-rate of the heavier dispersed phase. However, the particles to be sampled must not be so small that they follow the gas stream-lines and the mouth of the probe should be large, compared with the particle size. Reports on the use of sampling probes are given by Alexander and Coldren (1951), Longwell and Weiss (1953), Gill et al (1962) (1963), Janssen and Schraub (1968), Burick et al (1971) and Zeuker (1972).

A technique which allows simultaneous determination of the mass flows of both phases, and which overcomes the problem of the lower limit of particle size, is that of the isokinetic probe. A typical isokinetic probe is illustrated in Figure 6.11. The sampling rate is adjusted so that the velocities of the two phases, at the mouth of

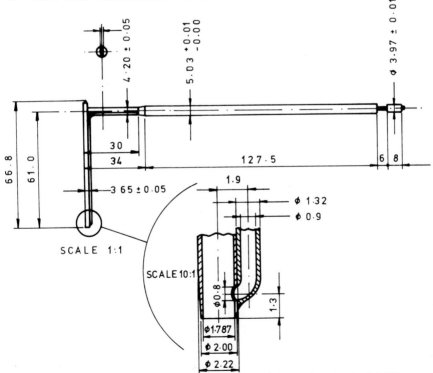

Figure 6.11: Isokinetic sampling probe used by Alia et al (1968).

the probe, are equal to what they would have been in the absence of the probe. This is achieved by having a pressure tapping at the mouth, which measures the difference between the pressure in the mouth of the probe and the static pressure in the flow channel. The flow is adjusted until the two pressures become equal (ie there is no impact pressure). Small corrections usually have to be made, to account for the fact that the pressure tapping on the probe cannot be exactly at the mouth. The problems of operation of isokinetic probes have been discussed by Parker (1968) and by Schraub (1969).

Examples of the application of isokinetic probe techniques, in gas-liquid flows, are the work of Alia et al (1968), Schraub (1966) and Kirillov et al (1978). Problems can occur where the sizes of the elements of a phase flowing in the channel, are equal to, or larger than, the size of the probe. This was illustrated in the work of Shires and Riley (1966), who attempted to apply isokinetic probe methods to bubble flow, where the bubbles were of greater size than the probe. This implies the use of quite large probes for many two-phase flows and an example of such a large probe ($\frac{3}{8}$ inch in diameter) is that of Burick (1972).

The isokinetic probe technique has been used extensively in gas-solid flow systems (Stukel and Soo (1968-69), Breugel (1969-70), Jepsen and Ralph (1969-70) and Boothroyd (1967a)). Boothroyd (1967a) describes the use of an alternative to pressure difference method, for the determination of the isokinetic condition. A hot-wire anemometer was placed immediately upstream of the probe and the flow in the probe adjusted, until the hot wire recorded the same value as it had in the absence of the probe.

Adorni et al (1963) described a special scoop-type probe for isokinetic mass-flow sampling, in the region immediately near the wall of the channel. This is illustrated in Figure 12. A very similar device has more recently been used by Dsarasov et al (1974).

An example of the results obtained, using isokinetic sampling probes of the types illustrated in Figures 6.10 and 6.11, is shown in Figure 6.13, where results are plotted for the same mass flow and pressure, but with alcohol and water, respectively, as the liquid phase. The differences in the liquid mass-flow, for the two liquids, are almost certainly due to the effect of surface tension.

The isokinetic principle can be applied to the measurement of sub-channel flows in axial flow through multi-rod bundles such as those used in nuclear reactor fuel elements. Here, the probe occupies the whole of the sub-channel and a pressure difference measurement is made between the entrance to the probe and the surrounding sub-channels. When this pressure difference is zero, the

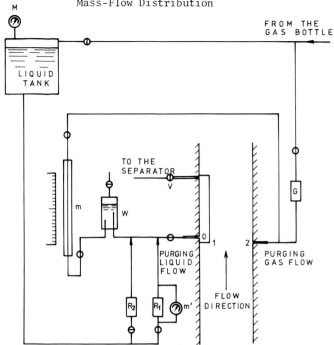

Figure 6.12: Wall scoop-probe arrangement used by Adorni et al (1963).

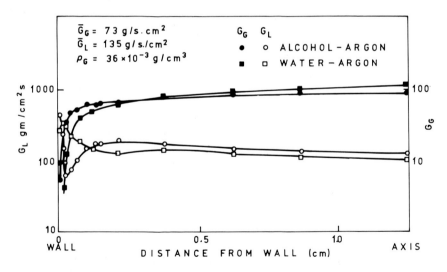

Figure 6.13: Mass-flow distribution in gas-liquid flow (Alia et al (1968)).

flow into the sub-channel probe is considered isokinetic. Measurements of this type are reported by Lahey et al

(1971), Castellana and Casterline (1972) and Bayoumi et al (1976).

An alternative form of probe for simultaneous flow rate measurement of the two phases is described by Burick et al (1974) and is illustrated in Figure 6.14. Here, the liquid passes into the probe but the gas is diverted, generating its dynamic pressure at the probe throat. The droplets decelerate down the probe and are collected and their flux measured. The over-pressure at the throat is mainly due to the dynamic pressure of the gas but it is possible to make a correction for the liquid contribution due to droplet deceleration between the probe tip and the point of gas pressure measurement. Excellent results were obtained with this probe and it seems worthy of wider application.

Good results can be obtained using isokinetic probes, provided attention is paid to problems such as those of keeping the pressure-tapping lines clear.

The development of isokinetic and related probes for transient (rapidly time-varying) flows would appear to be a promising line of investigation.

Phase Concentration-Distribution (Local Void Fraction)

In two-phase flow, the phases are rarely, if ever, distributed uniformly over the channel, nor are they split

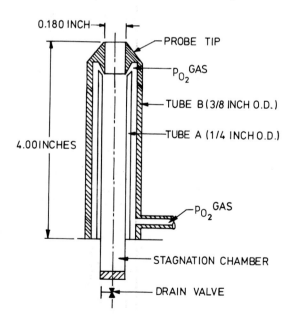

Figure 6.14: Concentric two-phase impact probe used by Burick et al (1974).

into two distinct regions as is often convenient to assume in separated flow models. The phase distribution is of interest in understanding the flows and, together with data on mass-flow distribution, can be used to make more accurate estimates of the average properties of the flow. The main methods which have so far been used for measuring phase distribution are now described.

The first method is that of gamma-ray absorption. By using a finely colimmated beam and traversing it across the channel, it is possible, by suitable (and often difficult), interpretation of the results, to deduce the diametrical distribution of void fraction. In round tubes, this type of measurement is often called "chordal mean measurement". Examples of its use are given by Kazin (1964), Staub and Zuber (1964), Kowalczewski (1966), Bolotov et al (1967) and Staub et al (1968). The interpretation of such measurements is discussed by Dementer and Skachek (1973). The problem of interpretation of the data is eased, if rectangular channels of high aspect-ratio are used and the measurement is made parallel to the longest faces of the channel. Such measurements are reported by Martin (1969), who gives a very large amount of data for boiling in both the sub-cooled and quality regions. Alternatively, collimated gamma scattering can be employed (Zielke et al (1975)); this technique was reviewed in Chapter 3.

An alternative method for obtaining void distribution is to use multiple beams from a single source. The principle is similar to that used by Smith (1975) and others for determination of flow pattern in horizontal tubes (see the first section of this chapter). The technique was used by Nylund et al (1968) in determining void fraction distributions in forced convective boiling in a 36-rod bundle. If a symmetrical distribution of void fractions can be assumed, then the results can be interpreted to actually give the distribution. Figure 6.15 shows some data obtained in these tests. The interpretation of multi-beam gamma densitometer data is discussed by Lassahn (1975) (1977); in the case of a single tube, the non-symmetrical distribution can be determined. For instance, with a two beam gamma densitometer, a void distribution of the form:

$$\rho(y) = \rho_L - \frac{\rho_L - \rho_G}{1+e^{-4a(y-b)}} \quad (6.2)$$

is assumed where ρ varies only with distance y from the bottom of the tube, then the two unknown parameters a and b can be determined from the readings of the two beams and the density distribution estimated. With three beams, it is possible to use a more complex distribution function (Lassahn (1977)).

By using a needle-contact probe which faces into the

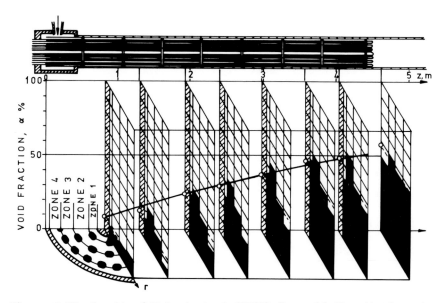

Figure 6.15: Result of Nylund et al (1968) for void distributions in a 36-rod bundle. (Pressure: 49.8 bar. Inlet subcooling: 3.7°C, Exit quality: 4.7%).

flow, it is possible to detect the time which the probe spends in the respective phases. This kind of probe is particularly suitable for bubble flow and an early development of its use was for liquid-metal systems (Neal and Bankoff (1963), Chen et al (1968)). Although earlier attempts to use the method, with water, indicated difficulty due to the hysteresis in the contact of the needle with the water phase (Nassos (1963), Nassos and Bankoff (1967)), other work with the system has proved very successful (Lackme (1964) (1965) (1967), Nassos (1965), Delhaye and Chevrier (1966), Malnes (1966), Sekoguchi (1968), Roumy (1969), Iida and Kobayasi (1970), Herringe (1971), Sandervag (1971), Gustafsson and Kjellen (1971), Lecroart and Lewi (1972), Uga et al (1972), Lewi (1973), Kobayasi (1974), Herringe and Davis (1974), Sekoguchi et al (1974), Michiyoshi et al (1974), Rubin and Roizen (1974), Sekoguchi et al (1975), Serizawa et al (1975), John et al (1976), Michiyoshi et al (1977), Mori et al (1977), Subbotin et al (1977), Adler (1977) and Reiman and John (1978)). Applications of the probe have included in-pile measurements (Uga et al (1972)) and measurement in liquid-liquid systems (Hoffer and Resnick (1975)).

A thorough investigation of the response of resistivity probes, for local void-fraction measurement, is reported by Delhaye and Chevrier (1966). As the probe approaches the bubble wall, a liquid film is formed round the probe

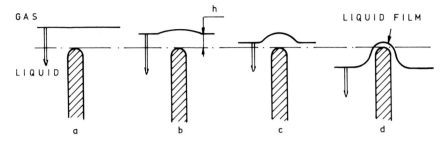

Figure 6.16: Distortion of liquid surface by probe tip, leading to contact hysteresis (Delhaye and Chevrier (1966)).

tip as illustrated in Figure 6.16. The probe thus shows a conductivity response which departs from the ideal, as illustrated in Figure 6.17. The time lapse for the removal of the liquid film from the probe tip (τ) varies with the type of probe and with the velocity of the fluid, as illustrated in Figure 6.18. Obviously, the probe can not be used if τ is greater than the bubble contact time. By electronic signal processing, it is possible to detect points A' and C' but small errors are still possible, as illustrated in Figure 6.17. Delhaye and Chevrier make the very important point that the method should not be employed for small bubbles with high velocity, due to separation of the bubbles in the liquid stream-lines around the probes. Typical errors in void fraction, measured by this means, are of the order of 20% and a typical set of results obtained by the method is illustrated in Figure 6.19. Note, in this latter figure, the peaks in void

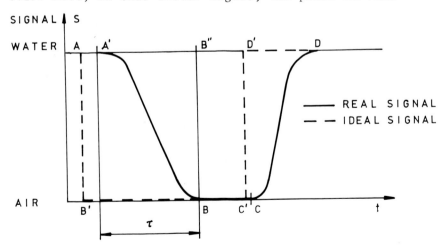

Figure 6.17: Differences between ideal and actual signals from needle contact probe (Delhaye and Chevrier (1966)).

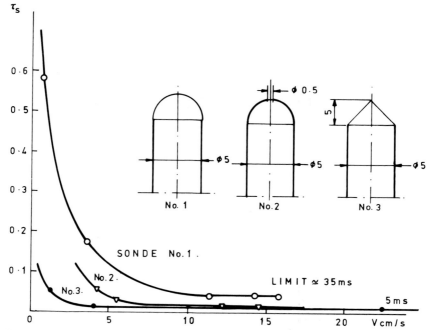

Figure 6.18: De-wetting delay times as a function of bubble velocity in the needle contact method (Delhaye and Chevrier (1966)).

fraction obtained in the region near the wall. This behaviour occurs at low liquid velocities and an interesting investigation of individual bubble motion by Sato (1974) may provide the explanation. At low velocities,

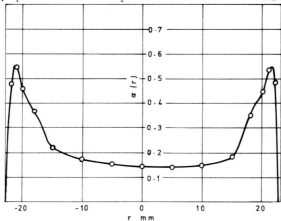

Figure 6.19: Diametrical distribution of void fraction in air-water flow. Mean liquid velocity, 1.5m/s, mean void fraction, 0.283. (Malnes (1966)).

bubbles injected near the wall stayed there and bubbles injected in the centre of the stream migrated to the wall flowing close to it with a "bouncing" motion. At high velocities, and for all velocities in downward flow, the bubbles tended to migrate towards the centre of the tube.

The principle of the hot-wire or hot-film anemometer probe is similar to that of the needle-contact method and relies on differing responses in gas and liquid. However, the response of the probe does not rely on the liquid being conducting. The use of hot-wire anemometers, in two-phase flow, is described by Hsu et al (1963) and further measurements are reported by Goldschmidt (1965), Goldschmidt and Eskinazi (1966), Delhaye (1968a) (1968b) (1969) and Dix (1971), Shiralkar and Lahey (1972) and Jones and Zuber (1977). A histogram showing probability of signal level as a function of signal level shows two peaks, one representing the liquid-phase flow and the other the gas flow. The void fraction can be obtained by directly integrating the areas in the histogram occurring for the two regions.

Optical probes of small dimension have been developed for local void-fraction measurement by Miller and Mitchie (1969) (1970), Danel and Delhaye (1971), Hinata (1972) and Delhaye and Galaup (1974). The Miller and Mitchie probe, illustrated in Figure 6.20, uses the principle that incident light, passing down the glass rod, is totally reflected and returned when the probe is in the gas phase. An even smaller probe made from a glass fibre is described by Danel and Delhaye (1971) and Delhaye and Galaup (1974) and is illustrated in Figure 6.21. More recently, a design successfully combining features of several of the earlier designs has been described by

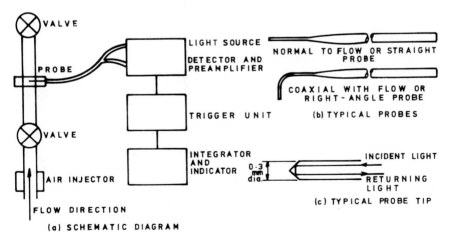

Figure 6.20: Optical probe for local void-fraction measurement in two-phase flow (Miller and Mitchie (1970)).

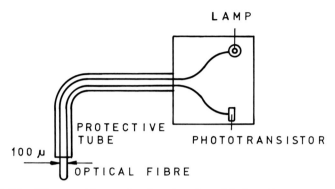

Figure 6.21: Fibre-optic probe for local void-fraction measurement (Danel and Delhaye (1971)).

Abuaf et al (1978). This latter probe is illustrated in Figure 6.22. A probe of similar principle to that of Miller and Mitchie has been used for liquid depth gauging and is described by Geake and Smalley (1975). In principle, optical probes can be applied to any fluid system and the signal processing is rather similar to that for needle-contact probes.

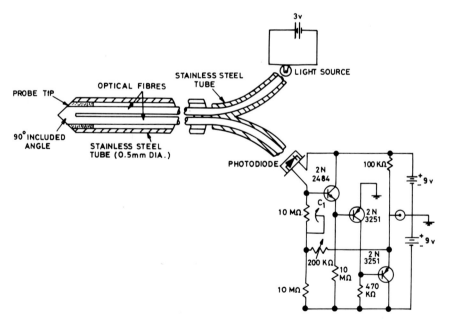

Figure 6.22: Optical probe used by Abuaf and Jones (1977) and Abuaf et al (1978).

Lafferty and Hammitt (1967) report the use of a two-wire conductance probe, for the measurement of void fraction. This probe can be traversed across the flow to measure local void-fraction and relies on conduction through the liquid phase. This method would be expected to show considerable sensitivity to the regime of flow.

Light scattering techniques have been used by Becker et al (1967) and by Danel (1971) to measure the concentration of a dispersed phase. Alternatively, the photo extinction of an incident beam may be measured and a technique for photo-extinction measurements, using fibre-optic devices, has been reported by Soo et al (1964) and by Stukel and Soo (1968-69). There are difficulties in maintaining clean conditions at the probe tips and a photo extinction device using an air purge system to avoid this problem is described by Fenton and Stukel (1976).

The use of isokinetic probes was described above. Such probes measure the local mass-fluxes but can also be used to measure an impact pressure, in the absence of extraction through the probe. This gives a third measurement from which, through a suitable model, the local phase concentration can be deduced. Such an approach has been adopted by workers at CISE, Milan (Adorni et al (1961a) (1961b) (1963) and Alia et al (1968). Adorni et al (1961a) present a theoretical model for impact pressure, in which it was assumed that the gas phases gradually diverted around the probe (as in a normal Pitot tube), whereas the liquid is assumed to be stopped, relatively suddenly, just before the probe mouth. Thus, the liquid is assumed to impart its total momentum to the probe and, provided the liquid and gas mass-flows are known from separate isokinetic measurements, it is possible to estimate the void fraction. However, examination of the data obtained by CISE for local void-fraction, shows that the slip ratios, even in fully developed annular two-phase flow, and near the centre of the channel, are of the order of 2-3. That is, the gas is apparently travelling at three times the velocity of the liquid. Naturally, there would be slip, but this is likely to be small, as is confirmed by calculation and in experiments by Cousins and Hewitt (1968a). Thus, the high slip-ratio, as reported by Adorni et al (1961a), must be regarded as being due to too high an impact force being assigned to the liquid phase. It seems probable that the impact pressure can be calculated reasonably accurately from a homogeneous model representation of the gas core.

No one method can be recommended for all cases of phase-concentration distribution measurement. Chordal mean void-fraction measurements are interesting, but often difficult to interpret. Best results are obtained in favourable geometries, such as rectangular channels with high aspect-ratio. For bubble-flow systems, the needle-contact probe has reached a high state of development and the point contact principle of measurement can be extended by using hot-wire and hot-film anemometry probes and, possibly preferably, optical probes. Light transmittence and scattering methods can be used in some circumstances, but methods relying on impact pressure measurement should be treated with reserve.

Further development of the gamma scattering method would seem to be worthwhile. There would also seem to be scope for the continued development of improved optical techniques, particularly if they can be combined with simultaneous velocity measurement (see below).

Velocity and Momentum-Flux Distribution

If the local phase concentration and the local mass-flux are known, the velocity can be calculated for each phase. Discussion of mass flow and phase concentration distributions is given in earlier sections of this Chapter. However, much work has been carried out on measurement of local velocity, by more direct means, and this work is reviewed in this section.

When the phases are completely separated, the velocity distribution in each separated zone can be determined using conventional methods such as Pitot tubes and hot wires. Information on single-phase flow measurement using hot-wire anemometers, is given in, for instance, the papers by Kidron (1967), Resch (1968), Gjessing et al (1969), Bruun (1971), Artt and Brown (1971), Freymuth (1972), Dent and Derham (1972), Downing (1972), Morrison (1974), Wyler (1974), Arrowsmith and Foster (1974), Morrow and Kline (1974), Lord (1974) and Firasat (1975). Other single-phase techniques, which might find application in two-phase flows, are ultrasonic flowmetering (Torrance (1973), McShane (1974) and Lynnworth (1975)), nuclear magnetic-resonance flowmeters (Vander-Heyden and Toschik (1971) (1975)), vortex-shedding flow-meters (White (1974)) and multihole Pitot probes (Walters et al (1971)). Pitot measurements on annular flow without liquid entrainment, and on horizontal separated flow in a rectangular duct, have been reported by Pike (1965) and Wurz (1971). Cases where complete separation occurs are rare and, more usually, a Pitot probe or other device would see both phases, either continuously or intermittently. For dispersed flows, Pitot tubes have been used, coupled with an assumption that the flow behaves as a homogeneous mixture; examples of the use of Pitot probes, in gas-liquid flows, are the work of Gill et al (1963a) (1963b), Delhaye (1966), Dzakowick and Dix (1969), Samoilovich and Yablokov (1970), Sterlini and Trotignon (1971), Burick et al (1971), Young et al (1974) and Banerjee et al (1978). The technique has been used in gas-solids flow by Dussourd and Shapiro (1955) and Dalmon and Lowe (1957). Both in the work of Gill et al and that of Dalmon and Lowe, checks were made on the integration of the total phase flows across the channel and these gave reasonably good agreement with the input flows. Typical results obtained by Gill et al (1963a) (1963b) are illustrated in Figure 6.23.

A theoretical study of the response of Pitot probes in droplet/gas flows is given by Crane and Moore (1972). They calculate that the homogeneous assumption can lead to a typical underestimate of the velocity by 1-2% for low-pressure steam conditions. Careful purging of the probe, by an air-stream, is necessary in order to achieve accurate measurements.

Using an isokinetic probe, if an assumption of homogeneous flow is made, the velocities of the phases are readily calculated. In this sense, the isokinetic probe is as efficient as a Pitot probe, which also depends on the homogeneous flow assumption. For local slip between the phases, a further variable needs to be determined and one possibility is to measure, directly, the local momentum-flux. The development of two-phase momentum-isokinetic probes has been described by Ryley and Kirkman (1967-68) and by Rao and Dukler (1971), the devices used being shown in Figure 6.24. Ryley and Kirkman used a momentum cage within the probe itself and Rao and Dukler used an isokinetic probe suspended on a beam. In the latter case, the

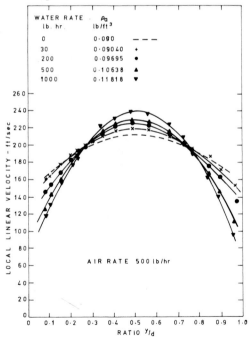

Figure 6.23: Velocity profiles determined by the Pitot-probe method, for air-water upwards annular flow in a tube (Gill et al (1963)).

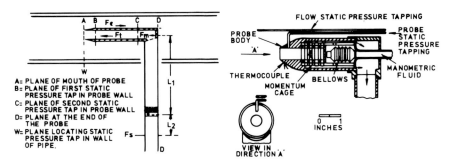

Figure 6.24: Isokinetic-momentum probes for two-phase flow studies.

force due to the forward momentum of the gas-solid flow entering the probe, plus the aerodynamic force on the probe, gave a deflection which could be measured. The aerodynamic force was measured by reversing the probe so that it pointed away from the flow direction, so that only the aerodynamic force was felt. The Rao and Dukler method has the advantage over that of Ryley and Kirkman that the probe is very much smaller. A momentum-flux cage device is also described by Baumeister et al (1973).

Two-phase momentum flux can also be measured using drag-disc devices, these are used in single-phase flow. Development work on the use of drag discs in two-phase flow is described by Aerojet Nuclear Company (1974) and by Fincke et al (1978). A typical unit used in their work is illustrated in Figure 6.25. It should be noted that measurement of two-phase momentum flux, coupled with density and turbine measurements, is being used as a means of measuring fluid mass-flow in the LOFT reactor safety tests. Further details of these applications are given in the sections on mass flow measurement in this Chapter and in Chapter 4. However, some assumption about the nature of the flow has to be made, in order to carry out this calculation and careful attention to calibration is necessary. This point is well demonstrated in the results listed earlier.

An investigation of a "momentum vector sensor" is reported by Upson (1974). The sensor consists of a 'drag rod' which passes through a pivot gland and protrudes into the channel carrying the fluid. Forces on the rod can be detected in both directions and magnitude. The device is useful in detecting flow direction but is less accurate than, say, the drag disc in measuring a wide range of fluid velocity. Papadatos and Svrcek (1974) report the use of force measurements on a pipe bend to measure pulverised solids flow.

Photographic techniques have been widely used for measuring local velocities. In these techniques, either the motion of individual elements of a phase (eg a bubble or a droplet) are photographed and measured, or tracer

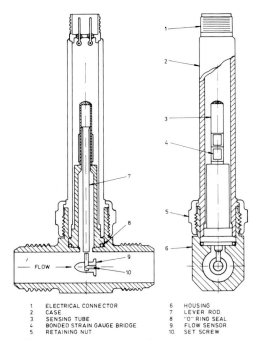

1	ELECTRICAL CONNECTOR	6	HOUSING
2	CASE	7	LEVER ROD
3	SENSING TUBE	8	"O" RING SEAL
4	BONDED STRAIN GAUGE BRIDGE	9	FLOW SENSOR
5	RETAINING NUT	10	SET SCREW

Figure 6.25: Drag-disc flowmeter. Aerojet Nuclear Company (1974).

particles are deliberately introduced into the flow, which are characteristic of the velocity in the locality. Discussions of methods of formation of tracer particles are given by Griffin and Votaw (1973), Nychas et al (1973) and Komasawa et al (1974). The use of tracer particles for measuring film surface velocities, is reported by Davis and Cooper (1969).

An extension of this technique, using laser scattering in the determination of surface particle velocities, is described by Blows and Tanner (1974). Measurement of velocity profiles in liquid films using tracer particles, is described by Persson (1964), who measured velocity by determining particle track length in a given time. If the particle concentration is uniform then the velocity profile can be determined by dividing the film up into arbitrary layers. The fastest particles are in the outer layer, the next fastest in the next layer, etc. The number of particles in each layer is equal. In most applications, however, the flow is multi-dimensional and stereoscopic photographic techniques are necessary in order to determine the components of local velocity. Stereoscopic methods, for liquid film flows, have been described by Nedderman (1961), Wilkes and Nedderman (1962), Cook and Clarke (1971), Azzopardi and Lacey (1974) and Azzopardi (1977). Application of stereoscopic photographic methods, for gas-solid flows, is described by Reddy et al (1969). These authors concluded that particles, even as large as 200 microns in diameter, followed the turbulence of the gas phase and travelled at virtually the same velocity as the gas. Other measurements on gas-solid flows are reported by Wakstein (1966). The development of holographic methods has proved a powerful alternative for studying three-dimensional motion. Holographic studies in two-phase flows are reported by Matkin (1968), Webster (1971), Lee (1973) and Lee et al (1974). Photographic methods are generally limited to cases where the concentration of the dispersed phase (bubbles, particles or droplets) is low.

A number of techniques have been evolved for the local addition of tracers, giving a highly-localised flow visualisation and the possibility of detailed local velocity measurements. Examples of these methods are as follows:
(1) Electrolytic method. Baker (1966) and Gerrard (1971) describe a technique in which water is doped with thymol blue and titrated almost to its end point. The water flows over electrodes, one of which is a fine wire located in the region of interest. A pulse of current produces a "cylinder" of dye tracer from the wire. Several such electrodes allow tracers to show three-dimensional flow patterns.
(2) Heat labelling. Here, a pulse of heat, from a heater or from a source of infra-red, is fed into the system and the hot zone produced observed using thermometry or fluorescent photography. A review of thermal flow measuring devices is given by Benson (1974). Parker (1965) carried out experiments with a foil temperature-sensor, in the form of a grid. This sensor was mounted on a metallic substrate and was subjected to step changes in temperature. Its response was characterised by a simple exponential curve with a single time constant. Hoffman de Visme and Singh (1972) derived an analytical relation for the thermal flow-meter, in terms of heat-injection rate and mass flow-rate over the laminar and turbulent range. Confirmatory tests, on a thermal flow-meter, show that pulsed heat input gave an improved output, compared with a DC-heated source, and the output signal depended directly on the flow rate. Velocity can also be obtained by measuring the fluid transit times from cross-correlations of local-temperature fluctuations at two points in the channel (Randall (1969) (1970), Termaat (1970)). A form of transit-time flow-meter was used by Pekrul (1970) to determine the flow-velocity characteristics and diffusion of thermal patterns. His sensors each employed an array of 10 thermocouples mounted on a stream-lined support, across the tube diameter. Roughton and Crosse (1974) measured flow by heating a small area of the wall of the channel which carried

flow and then studied the time decay of wall temperature, as the fluid cooled the wall. An accuracy of 2-3% was claimed, with calibrations, and 15-20% without calibrations.
(3) Flash photolysis. This technique is described by Popovich and Hummel (1967). The liquid contains a very dilute solution of 2-(2,4-dinitro-benzyl)-pyridine. A blue line is formed in the liquid (ethyl alcohol) on exposure to a high intensity, collimated beam of light. Subsequent motion of the line of tracer can be observed and the velocity profile deduced from it.

An alternative form of tracer technique, is to use radioactive tracer elements and the motion of tracers can be determined from time of flight along the channel, using a bank of detectors. This method has been used for gas-solid flow, in which one of the solid particles was tagged, by Gauvin (1964) and Toda et al (1973). It has been used for gas-liquid bubbly flow by Campanile et al (1969), who injected a radioactive bubble into the flow, to measure the bubble velocity. The use of soluble (and dispersible) tracers was discussed briefly in Chapter 2 above, in the context of quality metering. A general discussion of dispersible tracer techniques, for velocity measurement, is given in the papers of Clayton (1964) and Lafferty (1971). The use of tracers in air-water flows, in measuring the mean velocity of the liquid phase, is discussed by Evans (1970), who shows that considerable errors can occur, as a result of the time delays in thorough mixing of the phases across the channel. A further discussion of tracer techniques is given in the section on mass flow measurement earlier in this Chapter.

The needle-contact method, described in the context of void fraction above, can be adapted to the measurement of local phase-velocity, by having two probes displaced from one another in the line of flow and measuring the time delay between them, as the bubble passes from one probe to another. This method has been used by Neal and Bankoff (1965), Magrini (1966), Lackme (1967), Sekoguchi (1968), Lecroart and Porte (1971), Kitayama (1972), Reocreux and Flamand (1972), Lecroart and Lewi (1972), Lewi (1973), Kobayashi (1974), Michiyoshi et al (1974), Serizawa (1974), Sekoguchi et al (1975), Serizawa et al (1975), Mori et al (1977) and Michiyoshi et al (1977). The probe design used by Serizawa is illustrated in Figure 6.26. Using this probe, coupled with fairly sophisticated electronic processing, Serizawa obtained bubble-velocity spectra, of the type illustrated in Figure 6.27; average velocity data and standard deviation of bubble velocity can be obtained from these data and is exemplified by the results shown in Figure 6.28.

The optical probe technique, also described above in the context of void distribution measurement, can also be used for local velocity measurement. Again, a double

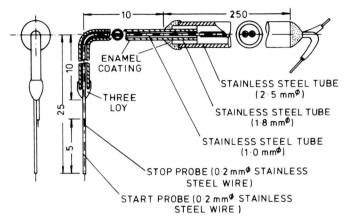

Figure 6.26: Double needle-contact probe for bubble-velocity measurement (Serizawa (1974)).

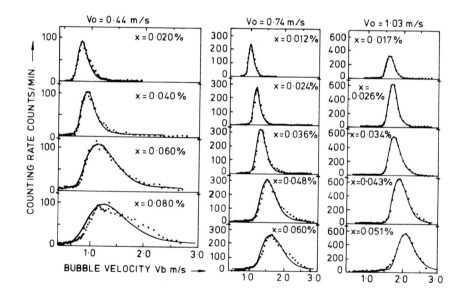

Figure 6.27: Bubble-velocity spectra determined using the double-needle-contact method (Serizawa (1974)).

probe is used operating on the same principle as the double resistivity probe described above. A double optical probe, consisting of two probes of the type illustrated in Figure 6.21 displaced in the direction of flow, is described by Galaup and Delhaye (1976) and the development

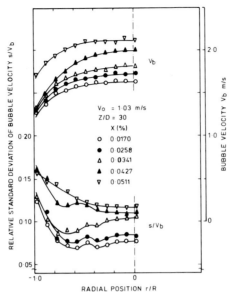

Figure 6.28: Profiles of the mean and standard deviation of bubble-velocity; air-water bubbly flow in a tube (Serizawa (1974)).

of new forms of double optical probe is described by Jones (1977); the probe used by Jones is shown in Figure 6.29.

An electrolytic probe for local liquid velocity measurement is described by Lorenzi and Pisoni (1973).

Over recent years, rapid development of the laser anemometry technique has been observed. A number of different systems have been developed and reviews of the subject area are given by Harris (1970), Stevenson and Thompson (1972) and Durst et al (1976). Typical of the applications of the technique to single-phase flow are those reported by Pike et al (1968), Morton and Clark (1971), Durst and Whitelaw (1971), Bedi (1971), Adrian and Goldstein (1971), Blake et al (1971) and Fitzgerald (1974). The two main variations of the technqiue are illustrated in Figures 6.30 and 6.31 respectively. The theory of these two methods is discussed by Wang (1972). In the first (Figure 6.30 (eg Iten and Mastner (1974)) light scattered from a laser beam by particles moving with the fluid, is Doppler frequency-shifted in proportion to the velocity component parallel to the exterior bisector of the scattering angle. The frequency shift is small compared with the optical frequencies and is determined by mixing some unscattered light, from the same laser, with the scattered light, on a photodiode or photomultiplier. This mixing produces an ac component, at the shift frequency, in the electrical output from the detector, from which the velocity can be derived. In this

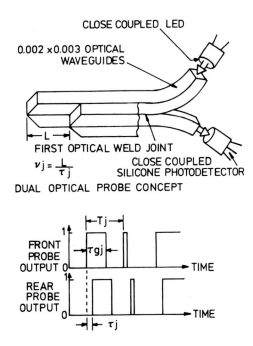

Figure 6.29: Double optical probe used by Jones (1977) for bubble velocity measurement in two phase flow.

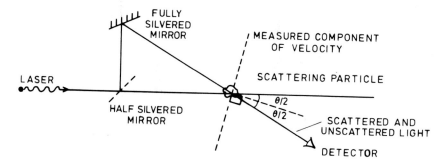

Figure 6.30: Doppler frequency-shift method for laser-anemometry measurement of local velocity.

variation of the technique, the component of velocity being measured, and its relationship to the frequency shift, are dependent on the positioning of the detector, relative to the incident beam.

In the second variation of the laser anemometry tech-

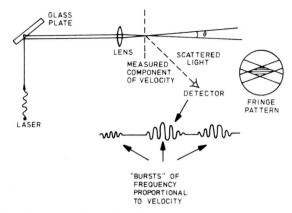

Figure 6.31: Fringe method for laser anemometry measurement of local velocity.

nique, (eg Mayo (1970), Bossel et al (1972)), two beams split from the same laser output are made to intersect, to produce an interference fringe pattern. Light scattered by the particles moving through the fringes is intensity-modulated, at a frequency in direct proportion to the velocity component normal to the fringes and in inverse proportion to their spacing. This modulation can be detected in light scattered in any direction from the fringes and produces an ac component, at the modulation frequency, in the output of a detector in any position relative to the beams.

The signal-to-noise ratio of this technique can be improved by polarising both of the crossing beams and using two detectors (Bossel et al (1972)). The cross-beam method can also be applied to the rapid measurement of velocity profiles; the crossing point is caused to scan the tube by means of a vibrating mirror (Bendick (1974)). The cross-beam technique can be extended to give flow direction sensitivity, by arranging for the interference fringe pattern to move, by shifting the frequency of one of the crossing beams. This variation is discussed, for instance, by Greated and Durrani (1971), Wilmshurst (1971) Oldengarm et al (1973) and Farmer and Hornkohl (1973), Sullivan and Ezekiel (1974) and Wigley and Hawkins (1978).

An alternative method of obtaining directional sensitivity is to use polarised beams and to examine the phase relationship of the scattered light (Dandliker and Iten (1974)). A further form of optical arrangement for laser-Doppler velocity measurement is described by Rizzo (1975); he uses an interferometer system in which the reference beam is scattered from a rotating disc, thus providing a simple method of frequency shifting. Ultrasonic methods for frequency shifting are described by Denham et al

(1975) and electro-optic devices by Foord et al (1974). A laser-doppler velocimeter capable of continuous instantaneous-velocity measurement is described by Avidor (1974); the frequency shift can be converted to a proportional change in fringe radius via an interferometer and this, in turn, can be converted to an analogue output of instantaneous velocity.

The laser anemometry technique is readily applicable to dispersed flows, where the dispersion is of relatively low concentration. Lehmann (1968) reports the measurement of droplet velocity and the technique was also used for droplet flows by Golovin et al (1971), Popper et al (1974), Durst and Zare (1975), Crane and Melling (1975), Durst and Zare (1976), Helmick (1977), Lee and Srinivasan (1977), Lee and Srinivasan (1978) and Srinivasan and Lee (1978). Lee and Einav (1971), Farmer (1972), Einav and Lee (1973), Birchenough and Mason (1976), Jurewicz et al (1977) and Raash and Umhauer (1977) report the use of the laser method for measurements of velocity in the flows of suspensions and solids in liquids. The technique is proving extremely powerful; for instance, Lee and Einav (1971) found it possible to examine the solid particle behaviour in the boundary layer. It was shown that the concentration of solids in the boundary layer was much lower than in the bulk flow. Simultaneous measurement of velocity and particle size is possible and this point will be returned to below.

The laser-anemometry technique has been used for measurements in liquid film flows by Amenitskii et al (1969) and by Oldengarm et al (1975). Here, it is possible to feed in the light beams from outside the tube and the necessity to cross an interface, with the consequential refraction and dispersion of the beams, is avoided. In many forms of bubbly flow, a sufficiently clear view of the interstices of the system is obtained to allow the use of the laser anemometry technique. Work on the applications of laser anemometry to bubbly flow is reported by Davies and Unger (1973), Ohba (1974) and Durst and Zare (1975). In all three studies, simultaneous measurements of local void-fraction were made. Studies of velocity profiles in liquid-liquid annular flow are reported by Shertok (1976).

For highly complex two-phase flows, it is unlikely that the laser technique can be successfully applied, due to the refraction and dispersion effects mentioned above. A discussion of some of the signal-processing problems involved is given by Rasmussen and Jensen (1971) and Lading (1971). One possibility, for overcoming the problems of scattering of the laser beams, would be to use suitable light guides.

A velocity measuring device which may find applications in two-phase flow systems is described by Sleath (1969). He used a glass-fibre probe which was placed parallel to the boundary surface and at right angles to the flow. Strain of the fibre gave a measure of the velocity and was recorded by a strain gauge and bridge system. This device could be employed for measurements close to void surfaces and has the advantage of being directionally sensitive.

When a light beam passes through a two-phase medium, it is modulated in a complex way, due to scattering resulting from the passage through the interfaces of elements of the respective phases. If the intensity of the beam leaving the system is recorded, then it fluctuates in a manner characteristic of the interruptions of its path by the phase elements. If a similar beam is passed through the system down-stream of the first beam, then it, too, will be modulated and, provided the distance between the two beams is not too great, then the modulations will be somewhat similar and their time displacement can be taken as characteristic of the modulating phase velocity. A simple application of this technique, is to the precise measurement of slug velocity and such measurements have been reported by Nicolitas and Murgatroyd (1968). The technique has also been applied to the measurement of droplet velocities from a nozzle by Tsiklauri et al (1970). Due to the turbulent nature of the flow, the actual output signal can vary, to some extent, from one light beam to the next and it is thus difficult to estimate the transit time. One way round this difficulty, is to use correlation methods. The cross-correlation function is obtained by multiplying the signal obtained from the first detector by the signal obtained from the second detector after a time delay τ (see Chapter 7). The product of the two signals as a function of τ is termed the "cross-correlation function" and reaches a peak at the most likely transit time between the two detectors. The principle of the method is illustrated in Figure 6.32. Applications of this technique to velocity measurements in gas-droplet flows are described by Matthes et al (1970) and by Deich et al (1974), to gas solid flows by Piplies (1972) and Oki et al (1977) and to bubbly flows by Riebold et al (1970) and Varadi et al (1977). The method may find some useful applications in transient studies and it could, in principle, be applied to high-pressure flows in steel tubes. A system designed for this purpose by Riebold et al (1970) is illustrated in Figure 6.33. A similar principle has been applied, in dispersed flows, by Thompson (1968) who measured the time taken for a particle to travel between two laser beams.

Velocity measurement by the cross correlation method can be achieved by the use of signals other than light beam modulation. Examples of the various alternative signals which can be used in this context are:
(a) Gamma beam absorption (Lassahn (1975b) (1976)).

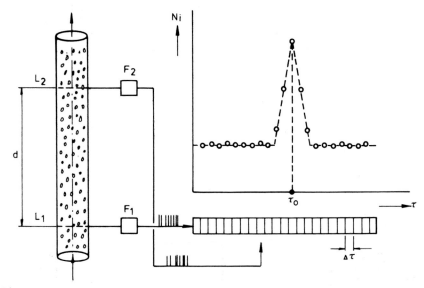

Figure 6.32: Illustration of the principle of the light-scattering cross-correlation technique, for velocity measurement in two-phase flow (Riebold et al (1970)).

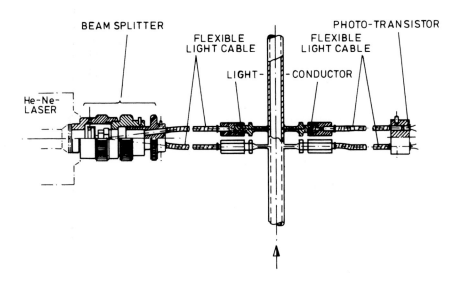

Figure 6.33: Design of light-scattering device for bubble-velocity measurement in steel tubes with high pressure steam-water flow (Riebold et al (1970)).

(b) Neutron noise fluctuations in reactors (Ando et al (1975), Kostic (1976) and Ceelan et al (1976)).
(c) Ultrasonic signals (Olszowski et al (1976)).

Measurement of local velocities in two-phase flows are interesting and important in understanding the flows. No one method can be generally applied, however, and the optimum selection of the method depends very much on the system to which it is to be applied. In bubbly flows, the double needle-contact method appears to offer the best advantages but the laser anemometry method can be applied in some cases. In liquid film flows, laser anemometry can be applied and is probably preferable to other techniques for film velocity measurements, such as stereoscopic photography. For more complex flows, the cross-correlation (light transmittence) methods may apply over limited ranges (though even they are unlikely to be useful for very high mass-flows). In the case of high velocities, with a complex flow-pattern, probably the best plan is to use isokinetic sampling, coupled with momentum-flux determination.

There is a strong need for improved methods for local velocity measurement in two-phase flows. Probably, these will come by the development of laser anemometry techniques when these are coupled with other instantaneous local measurements (eg void fraction or drop size) a wealth of new information about the flow becomes available.

Concentration Distribution

In many two-phase systems, mass transfer or simultaneous heat and mass transfer is taking place and this sets up concentration profiles. The determination of such profiles is of considerable interest in the interpretation of the mass transfer effects but few quantitative data exist in this area. The determination of concentration profiles, using probe techniques, is open to difficulties of interpretation. Some of the difficulties of suction probes, for the measurement of local concentrations in vapour-gas mixtures, are discussed by Navon and Fenn (1971). A novel probe, for the measurement of the concentration of solutes in the liquid phase flowing in a two-phase flow, is described by Jagota et al (1973c). This probe was used, in particular, in annular flow and had a porcelain tip which restricted the access of the gas phase into the probe, due to the high pressure required to pull the gas into the pores of the tip.

In some applications, it is possible to avoid the use of sampling probes and to use, as an alternative, interferometry techniques. Lin, Moulton and Putnam (1953) report the use of interferometry for measuring boundary layer concentration profiles in 'Concentration-polarised' electrolyte systems. Prasad et al (1971) described the use of interferometry in the study of heat, mass and momentum transfer from a horizontal air-water interface

The concentration and temperature profiles were examined, simultaneously, by using interferometry at two different wavelengths. Similar techniques for the studying of the evaporation of combusting droplets are reported by Ross and El-Wakil (1960). Other examples of interferometric measurements are those of Beer (1971), who studied the boiling of multi-component mixtures, Nakaike et al (1971), who studied interfacial turbulent effects in liquid-liquid systems, Hu and El-Wakil (1974), who studied n-heptane evaporation in a special cell. Recently, it has been possible to simplify and extend the interferometry techniques by means of laser holography. A qualitative study of this possibility is reported by Masliyah and Nguyen (1974) and an extensive review of the qualitative and quantitative applications of the methods given by Mayinger and Panknin (1974). In the classical application of interferometry, a light-beam passing through the test cell is compared with one passing through an exactly similar path, without the presence of heat or mass transfer. The problems of matching the paths are very considerable and great attention to detail is necessary for accurate results. However, it is possible to use a hologram of the actual test cell, in its initial condition without the transfer processes taking place, instead of using a reference cell. The principle is illustrated for the case of temperature distribution measurement in Figure 6.34. A plate is first exposed with a comparison wave and then, subsequently, with a wave passing through the test section with heat transfer taking place. On reconstructing the holographic image, an interference pattern is observed, indicating the temperature profile. The application of the technique to mass transfer alone is identical to that

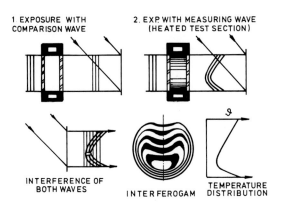

Figure 6.34: Principle of the double-exposure holographic technique for temperature or concentration measurement (Mayinger and Panknin (1974)).

for temperature measurement, but a powerful extension of the technique is that it can be used for simultaneous heat and mass transfer, by using two laser beams of different frequencies, as illustrated in Figure 6.35. Results reported by Mayinger et al for laminar boundary-layer temperature and concentration profiles, from a vertical heated plate, coated with a thin layer of naphthalene, with heat and mass transfer into a surrounding water bath, is illustrated in Figure 6.36.

A technique which is likely to assume increasing importance in the future for local and remote concentration and temperature measurement within fluids is that of

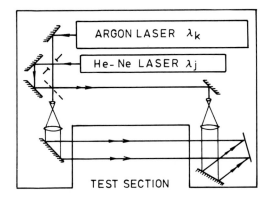

Figure 6.35: Two-beam method for simultaneous measurement of concentration and temperature profiles, by laser holographic interferometry (Mayinger and Panknin (1974)).

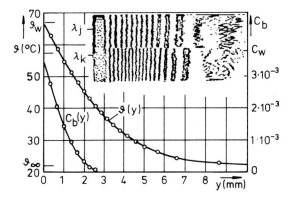

Figure 6.36: Temperature and concentration profiles in a laminar boundary layer determined by laser holographic interferometry.

laser Raman scattering. The principle of Raman scattering is illustrated in Figure 6.37. Suppose a fluid is irradiated with a laser beam of frequency ω_1; the beam interacts with the molecules of the substance exciting these molecules from the ground state, say, to a virtual state A (see Figure 6.35a).

The molecules then return from the virtual state A, re-emitting light; most of this re-emission is at the original frequency ω_1 but a proportion of the re-emission occurs due to the molecule returning from state A to a vibrational level, say, which has an effective frequency difference Δ from the ground state. The emission corresponding to this latter transition has a frequency $\omega_s = \omega_1 - \Delta$. The emission at frequency ω_s is called the "Stokes emission". Depending on the temperature, molecules already exist within the vibrational level and these themselves can be excited to a virtual state B by the laser irradiation (see Figure 6.37a). If there is a return from state B to the ground state, then emission occurs at the so-called anti-Stokes frequency $\omega_{as} = \omega_1 + \Delta$. By examining the scattering arising from a point along the laser beam length (within the fluid) and by determining the intensity of the scattering at ω_s or ω_{as}, it is possible to determine the local concentration of the molecular species. Since each molecular species has a characteristic series of frequencies, concentration of components within multicomponent mixtures can be evaluated locally. Since the population of the vibrational state is dependent on

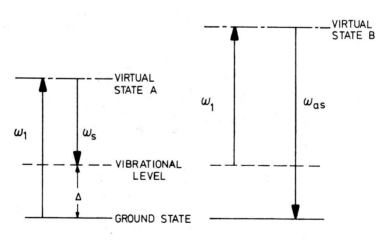

Figure 6.37: Principle of Raman scattering method for temperature measurement.

temperature, the local temperature of the fluid can be determined from the ratio of the intensities of the anti-Stokes and Stokes emissions.

Though it is a very powerful technique for remote detection of average local concentrations, the main problem with the conventional Raman technique is that the Raman scattering itself is very weak and long times (eg 10 minutes) are required to obtain accurate values. From the point of view of measuring two-phase flow systems, this would have made the Raman spectroscopy system of somewhat academic interest. However, a technique has been developed recently which allows measurements to be made (usually using pulsed lasers) in time of the order of nano-seconds. This technique is known as CARS (coherent anti-Stokes Raman scattering) and is illustrated schematically in Figure 6.38. A laser beam of frequency ω_1 is crossed with a beam from a tuneable dye laser at the Stokes frequency ω_S at the point where the measurement is required. This results in the creation of two *coherent* beams at the anti-Stokes and Stokes frequencies respectively which emerge from the system with an angular displacement relative to the incident beams as shown. The fact that these beams are coherent and separated from the incident beams means that highly rapid and high sensitive measurements can be made.

A review of laser Raman diagnostic techniques is given by Williams and Stenhouse (1978) and further detailed information is given in the papers of Regnier (1973), Begley et al (1974) Regnier et al (1974) and Nibler et al (1976).

Few quantitative data exist in the area of concentration distributions in two-phase flow. Probe techniques are difficult to interpret and more advanced techniques such

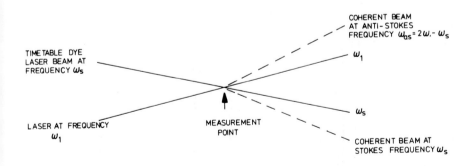

Figure 6.38: Coherent anti-Stokes Raman scattering (CARS) system.

as interferometry have limited application. Measurements of concentration distribution, except under very idealised conditions, are practically non-existent and techniques should be developed in this area.

Mixing Characteristics of Two-Phase Flow (including Phase Mass-Transfer)

Mass transfer of the phases, normal to the direction of flow, and the mixing of the phases are interesting in the development of theories for two-phase flow. For instance, in annular flow with droplet entrainment, the rate of transport of the droplets to the wall and the rate of creation of droplets from the wall film are of prime importance in calculating such flows in detail. Interphase interaction is also of interest and it is relevant to note, for instance, what the effect of a dispersed phase is, on the continuous-phase turbulence and whether or not the dispersed phase, itself, responds to the turbulent fluctuations in the continuous phase.

Problems of phase mass-transfer and mixing have been studied by a range of techniques, the main ones of which are now described.

The study of two-phase channel flow by tracer addition, involves the addition of a tracer at one point in the channel and the study of its axial or radial dispersion down-stream of the point of addition. This method has been discussed by Quandt (1962), Cousins et al (1965), Rosehart and Jagota (1978), Cousins and Hewitt (1968), Jagota (1970), Jagota et al (1973a) and Jagota et al (1973b) and Hills (1973), in the context of the study of droplet mass-transfer in annular two-phase flow. The principle of the method was used by Quandt is illustrated in Figure 6.39; a dyestuff is added to the liquid film in an equilibrium annular flow. The concentration of dye stuff in the film decays giving a measure of the interchange rate. Assuming complete mixing within the gas core and within the film, the change of film concentration with distance can be calculated from the equation:

$$\ln[(c_f - c_m)/c_m] = \ln\left(\frac{F_E}{1-F_E}\right) - \frac{\pi d_o I_z}{F_E(1-F_E)W_L}$$

where c_f is the local concentration in the film at distance z, c the mean concentration of dyestuff in the liquid, d_o the tube diameter, F_E the fraction of the liquid phase which is entrained as droplets and W_L the total liquid flow rate. The interchange rate, I, is equal to the deposition rate, D, and the entrainment rate, E, at equilibrium. Thus, a plot of $\ln[(c_f - c_m)/c_m]$ against z should be a straight line, I and F_E being determined from the slope and intercept. On investigating the Quant method, Cousins et al

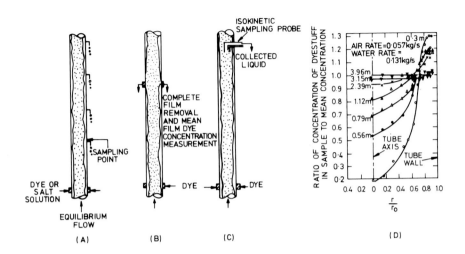

Figure 6.39: Tracer mixing methods for determination of film/core liquid interchange in annular flow.

(1965) found that the values of F_E estimated were very different to those measured by film suction. They determined the mean concentration in the liquid film by removing the whole film (Figure 6.39b) and found that this different greatly from the concentration measured at the wall, particularly in the region immediately above the dye injection point. Thus, they concluded that the mixing in the liquid film was rather slow and that this vitiated the Quant method. In a further study, Cousins and Hewitt (1968) used a modified dye transport method as illustrated in Figure 6.39c. The concentration profiles of dyestuff within the liquid phase were measured as a function of distance from the dye injection point and typical results are illustrated in Figure 6.39d. These measurements illustrate that any assumption of complete mixing within the gas core is open to question, It is possible to calculate, on the basis of the concentration profiles shown in Figure 6.39d, the mass transfer coefficient and this was found by Cousins and Hewitt to be of the same order as that measured in unidirectional deposition tests. However, the method is likely to be too tedious for widespread application. A possible new lease of life has been given to the dye transport method in recent work by Andreussi and Zanelli (1976) who used mean film concentration measurements and who showed, theoretically, that, though it is indeed true that the wall concentration may be very high, this high concentration zone is limited to the boundary sub-layer and the concentration in the rest of the liquid film is close to the mixed mean. However there may still be a problem due to the non-uniform concentration in the gas core as shown in Figure 6.39d.

Pulse injection of tracer in annular flow was used by Jagota et al (1972) and by Roukens de Lange (1975). The results throw interesting light on the interpretation of experimental observations in annular flow, in which pulses of radioactive tracer were injected and the transit time of the pulse measured. The work of Jagota et al shows that the most important mechanism, for axial dispersion, is the interchange of the droplets between the gas core and the liquid film. The effects of this interchange are illustrated in Figure 6.40. With increasing droplet interchange between the gas core and the liquid film, the pulse associated with the motion of the droplets is dispersed and decreased in velocity and the pulse associated with the liquid film is also dispersed and decreased in velocity. Thus, as the amount of droplet interchange

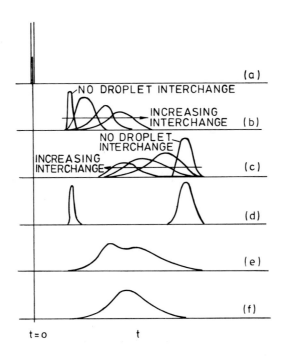

Figure 6.40: Movement of the tracer in the liquid phase of annular flow with entrained liquid.

is increased, the response at the measurement point develops from two separate pulses, corresponding to gas core and liquid film (Figure 6.40d)) to one, where there is a single pulse (Figure 6.40f)), travelling at approximately the mean liquid velocity. An analysis of tracer dispersion in annular flow is given by Stothart and Horrocks (1972).

The use of salt tracers in the study of phase mass-transfer, in steam-generating channels, is described by Styrikovich et al (1971).

The measurement of local concentration profiles, of a solute mixing in the continuous phase in dispersed flow situations, has been studied by Gomezplata and Brown (1968), Ohki and Inoue (1970), Subramanian and Tien (1975) and Stiegel and Shah (1977) for bubbly flow and by Boothroyd (1967a) and Haque and Boothroyd (1968-69) for gas-solids flow. Rowe and Evans (1974) made similar measurements in fluidised beds. Haque and Boothroyd concluded that the presence of the solid particles considerably reduced the eddy diffusivity in the continuous phase. Briller and Robinson (1969) measured the diffusion of particles, by using tracer particles in a gas-solid flow. They showed that the diffusion coefficients of the particles was lower than expected for the gas phase. This seems surprising, in view of the small diameter particles (1 micron) used. Russell and Lamb (1965) studied the dispersion in an annular film flow by a salt injection technique.

The problem of inter-sub-channel mixing is most important in predicting the behaviour of multi-rod fuel elements for nuclear reactors. Mixing studies are reported, for instance, by Bestenbreur and Spigt (1967), Du Bousquet (1968), Petrunik and St. Pierre (1970), Gonzalez-Santalo (1971), Castellana and Casterline (1971), Lahey et al (1971), Marinelli and Pastori (1972), Lorenz and Ginsberg (1973), Todreas (1974), van der Roos (1970), Gustafsson et al (1974), Mazzone (1974), Castellani et al (1975) and Mayinger et al (1977). Either one, or both phases, can be labelled in flow in one sub-channel and the transfer of tracer to the other sub-channel determined. A major difficulty applies in the interpretation of results obtained, using this technique for annular flow. The experimental models for the sub-channels are often in the form of two parallel channels, connected through a narrow gap, simulating the gap between adjacent fuel elements. In annular flow, however, one of the main mechanisms of transport might be around the film on the individual rods and this is not simulated in this kind of experiment.

If, in a dispersed flow of liquid droplets or solid particles, the suspended material migrates towards the channel wall, and if it is not re-entrained from the wall, then it is possible to measure the rate of transport, either by measuring the change in the flow rate in the gas core, or by detecting the amount deposited. Experiments of this type are not representative of, say, a fully-developed system in which there is equal entrainment and deposition, eg fully-developed annular flow. The difficulty in using data from this kind of experiment to represent annular flow is that, in the absence of a thick liquid film, the rate of deposition may be different from that which would occur with a highly roughened interface, characteristic of normal annular flow. Early experimenters on droplet deposition were Alexander and Coldren (1951) and Longwell and Weiss (1953), who used a sampling probe to determine the mass flow of droplets in the gas core. The rate of deposition was obtained by difference. In this kind of measurement, one problem is that the deposited liquid may form a new film, which may ultimately be re-entrained. Experiments on droplet deposition are reported by Cousins et al (1965) and Cousins and Hewitt (1968a). Fully developed annular flow was set up and then the film was removed through a porous suction device. The entrained droplets then began to deposit, forming a new film which was removed through a further extraction device, placed downstream of the first one. The liquid film flow, at the second extraction device, was less than that required to cause re-entrainment. Data obtained from these experiments showed that the droplet mass-transfer coefficient did not apparently depend strongly on the gas velocity. A somewhat similar technique was used by Ueda et al (1978). This technique has been extended for horizontal annular flow by Anderson and Russell (1968) (1970) and for flow over a flat plate by Simpson and Brolls (1974).

An experimental technique, to determine both the rate of deposition and the size of the droplets deposited, has been described by Farmer et al (1970), who injected a single stream of droplets into a gas flow in a channel with walls coated with magnesium oxide. Where the droplets collided, they formed a pit in the coating which could be seen from the outside of the transparent tube. The number of pits formed gave the rate of deposition and the size of the pits allowed an estimate of the droplet size. It was concluded from these experiments that the droplets closely followed the turbulence of the gas phase. Parker and Ryley (1969-70) studied aerosol deposition, by using particles of fluorescent material which could be washed off the surface subsequent to the experiment and their number estimated by measuring the concentration of the fluorescent dyestuff in the wash water.

For deposition of droplets in heated systems, an equivalent technique can be used, involving the measurement of burn-out. This technique was used by Hewitt et al (1969), for study of droplet deposition in steam-water flows at 70 bar and is illustrated in Figure 6.41. Burn-out, resulting from film dryout, is caused to occur simultaneously immediately up-stream of, and at the end of, a short length of tube at the end of a longer,

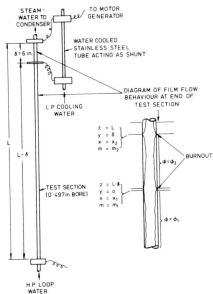

Figure 6.41: Heat-flux method for the determination of droplet deposition-rate in steam-water flows at high pressure (Hewitt et al (1969)).

uniformly heated, tube as illustrated. The short length at the end of the tube has a lower heat flux than the rest of the tube and this penultimate zone receives water only by deposition from the gas core. The deposition rate is equal and opposite to the evaporation rate in this zone and can be determined from the quotient of the heat flux and the latent heat.

The field of mixing and mass transfer would appear to warrant much further experimental investigation. Relatively little is known about phase motion in the direction perpendicular to the main flow axis and the subject is important in amny practical applications, particularly in the problem of extracting pollutants from outlet gases. Droplet mass-transfer is central in the modelling of annular flow for the prediction of burn-out, etc, (see, for example, Whalley et al (1974) and Hewitt (1978)). Particular attention needs to be paid to the uni-directional (droplet deposition/entrainment) rates of mass transfer in fully developed, developing and transient annular flows.

Drop, Bubble and Particle Size Measurements

In modelling the motion of fluids in a two-phase flow, containing a dispersed phase (droplets, particles or bubbles) suspended in a continuous phase (gas or liquid) it is necessary to have some idea of the size of the elements of the dispersed phase. Bubble size (or, more specifically, the gas-liquid interfacial area) is important in gas/liquid contactors and reactors and drop size is important in spray systems (including flames) and other dispersed flow systems. Reviews of interfacial area measurement methods are given by Alper (1976) and by Landau et al (1977). Surveys of drop size measurement techniques are given by Jones (1977) and Azzopardi (1977).

The main methods which have been used for measuring drop, bubble and particle sizes are photographic methods, impingement-coating methods, electrical-contact methods, light-scattering methods and laser anemometry related methods. A number of other techniques, not classified in this list, have been employed and are also briefly reviewed below.

Photographic Methods: The principle of the photographic technique is to take flash photographs, usually with a small depth of field, of the dispersed flow and to analyse them to obtain element size. Work of this kind, on droplet size, has been reported by Finlay and Welsh (1968). In many situations in which it is necessary to know droplet size, there is an annular liquid film which stops clear photography of the inner core. One way to overcome this is to remove the film and to photograph the core directly (Cooper et al (1963), Pogson et al (1970)). One difficulty in measuring small droplets is that of obtaining sufficient light intensity; this can be overcome by using lasers (Kirkman and Ryley (1969)). Another problem is that of identifying those particles which are in the plane of focus and it must be determined whether the particular object is a small, in-focus droplet or part of a large out-of-focus one. Making such decisions can be tedious and it is probably better to use holography, in which the 3-dimensional image can be reconstructed accurately after the photograph has been taken. Alternatively, drop sizes can be measured from axial view photographs (Whalley et al (1977), Mayinger et al (1976)) - see Chapter 7. The use of laser holographic methods for droplet size are discussed, for instance, by Fourney et al (1969), Webster (1971), Matkin (1968), Thompson (1974), Lee (1973), Bexon (1973), Lohmann and Shuman (1973), Lee et al (1974) and Webster et al (1976). Fast film analysing techniques for drop size are described by Simmons (1977).

Photographic determinations of bubble size in bubble-flow systems are reported, for instance, by Perkins and Westwater (1956). In bubbly flow, it is often impossible to see sufficiently far into the channel to be certain of the average bubble size. Kawecki et al (1967) have reported the use of sampling of bubles within a bubble column with subsequent photography. This technique raises doubts about the validity of sampling bubble flow without bubble coalescence.

Impingement Coating Technique: If a plate coated with soft magnesium oxide, (the coating is obtained by holding the plate over a burning magnesium ribbon), is held in a droplet flow, then holes are made in the magnesium oxide film by the impinging droplets. The size of the holes is slightly greater than that of the droplets. This technique has been used, for instance, by Ryley and Fallon (1966) and Kolb (1962). Farmer et al (1970) used this technique by coating the walls of a channel completely with magnesium oxide and noting the number, and size, or droplets deposited on the wall from a stream injected at the channel axis. Drop size measurements were made in annular flow by this technique by Ueda et al (1978); a special fast shutter was designed to take a representative sample of the drops.

Electrical Contacts Methods: If a droplet impinges on an electrically-charged wire, there is an instantaneous withdrawal of charge from the wire which can be recorded electronically. There have been many attempts to make this method work, of which the most successful reported is that of Gardiner (1964). For most systems, it is difficult to avoid the occurrence of spurious signals.

Wicks and Dukler (1966) have reported the successful use of a needle-contact device for drop size measurement. In this technique, two needles placed exactly in line, were brought together in the droplet stream and the number of times, for each spacing between the needles, at which the droplets bridged the gap, was counted electronically. By varying the gap size, it was possible to obtain the drop-size distribution. This technique is potentially powerful, since it allows in-situ measurement of droplet size, in annular flow with entrainment, and would also allow the radial distributions of drop size to be determined. This could give useful clues about the droplet mass-transfer. Further work, using the Wicks and Dukler technique, has been reported by MacVean and Wallis (1969) and Pye (1971). Experience with this system has not so far been very encouraging. In early trials of the method at Harwell, the count rate from the probe was found to be a function of the applied voltage and this was confirmed by MacVean and Wallis. Interpretation of the count-rate data, to obtain a droplet size distribution, is also open to error, as indicated by MacVean and Wallis.

The needle-contact method for measuring void fraction and velocity in bubble flow (described previously in this chapter) can also be interpreted to yield bubble size information. This is done by measuring the times between the first contact and the final contact for individual bubbles and also measuring the bubble velocities. The products of the difference in time and the velocity gives a distribution of the chordal distances passed through the bubbles by the probe and this, in turn, can be transformed to a bubble size distribution. Examples of the use of this technique are the work of Sekoguchi et al (1975) and Michiyoshi et al (1977).

Light Scattering Methods: A light beam passing through a dispersion will have an extinction proportional to the effective superficial area. This is true for both transparent dispersed phases and opaque ones. Provided the received beam is exactly in line with the transmitted one, the transparent dispersed elements will scatter the incident beam and behave as though they themselves are opaque with regard to the received beam. This technique of light scattering, to measure interfacial area (and, thus, element size), has been used for bubble flows (Calderbank (1958) and Lockett and Safekourdi (1977)), liquid-liquid emulsions (Langlois et al (1954)) and droplet suspensions (Walters (1969) (1971)). Reith (1970) has shown that the light-scattering technique gives results for interfacial area different from those obtained using chemical reaction methods. In using the attenuation method for bubble-or drop-size, it is, of course, necessary to know the concentration of the dispersed phase present and lack of information on the concentration often renders the technique inapplicable. Furthermore, for very small particles, (of the same order as, or less than, the wavelength of the light) the scattering is often partly in the direction of the incident beam and these particles will not contribute as much as expected to the total attenuation. However, the fact that the scattering behaviour is strongly influenced by particle size, can be used to good effect in the development of particle size-measurement methods, particularly for small particles.

For particles of size much less than the wavelength of the incident light, the classical scattering theory of Mie (1908) is applicable. For larger particles, the scattering characteristics are determined by geometrical optics methods and are discussed, for instance, by Davis (1955), Gucker and Rowell (1960), Gucker and Egan (1961) and Woodward (1964). Deich et al (1971) (1972) described the application of scattering techniques to the determination of particle size, in wet steam flows in nozzles. Their technique is based on the application of the Mie scattering theory and they measure the intensity, I_A, of the light scattered forward at an angle close to the direction of the incident beam and intensity, I_D, of the light scattered backwards at an angle close to the respect to the incident beam. According to the Mie theory, the ratio I_A/I_B is a unique function of the radius of the particles. It is possible to measure droplet sizes in the range, 1.2×10^{-8} m $< 8 \times 10^{-7}$ m. The maximum concentration of droplets should not be higher than $10^{14} - 10^{12}$ drops per cm^3 for the same range of droplet sizes. For drop or particle sizes outside the Mie scattering range, the scattering pattern is more complex. However, the intensity of a pulse of scattered light, emitted from a

given droplet or bubble, will still depend on its size
and the calibration, between size and pulse height
recorded, can be established by experimentation. The
scattered-light counting method shows considerable promise
in the investigation of nucleation of bubbles and this is,
of course, an important factor in the study of blow-down
transients. Work on the application of the technique to
bu-ble size measurement is reported by Landa et al (1970),
Landa and Tebay (1972) and Keller (1972). The apparatus
used by Keller is shown in Figure 6.42. Bubbles passing
through the control volume gave a pulse of scattered
light into the photo-multiplier, which was characteristic
of the bubble size. In order to calibrate the system,
Keller used electrolytically-generated bubbles, though
the minimum size which could be produced by this technique
was about 10 microns. For smaller sizes, a theoretical
estimate was made of the variation of scattering intensity
with bubble radius.

A method for the application of the scattering method
to the continuous determination of drop size distribution
was developed by Swithenbank et al (1976) and a commercial
unit using this technique is available (Malvern Instruments
Ltd, UK). The application of the method to dropsize
measurement in annular flow is described by Azzopardi et
al (1978) and a sketch of the system used is given in
Figure 6.43.

The liquid film is separated and the resultant droplet
stream passes between two windows through which the laser
beam passes. The beam is diffracted at a small forward
angle as illustrated. The detector consists of a series of
annular rings and the distribution of intensity over the
respective rings can be processed to give a dropsize
distribution. Data for drop size in annular flow are
illustrated in Figure 6.44 from which it will be seen that
dropsize is relatively insensitive to liquid mass flux,
though it increases rapidly with decreasing gas mass flux.

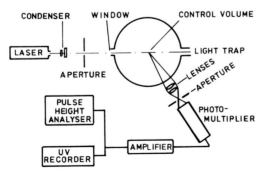

Figure 6.42: Apparatus for the light-scattering method for the
determination of cavitating bubbles (Keller (1972)).

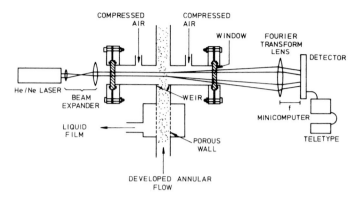

Figure 6.43: Defraction method for dropsize determination in annular flow. Azzopardi et al (1978).

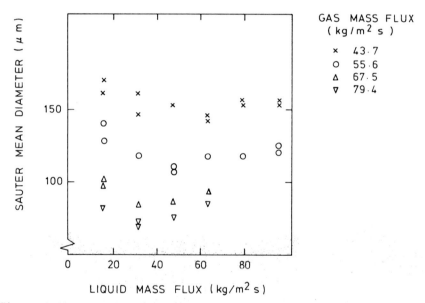

Figure 6.44: Dropsize distribution data for air-water annular flow. Azzopardi et al (1978).

Laser Anemometry Linked Methods

The amplitude of the "bursts" received in the signal from a laser anemometer (see Figure 6.31) is dependent on the particle size. Studies of drop size using this dependance are reported by Lee and Srinivasan (1977), Lee and Srinivasan (1978) and Srinivasan and Lee (1978). Figure 6.45 shows the calibration curve obtained by Lee and Srinivasan (1978) for monodisperse drops generated from an orifice. Use of laser dopper anemometry signals

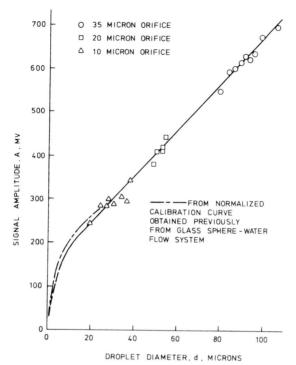

Figure 6.45: Relation between laser doppler anemometer burst amplitude and drop size (Lee and Srinivasan (1978)).

offers the very interesting possibility of making measurements not only of drop size distribution number density and velocity but also of the cross links between the variables. Thus, the number of droplet within a given range of axial and radial velocity can be determined by suitable electronic processing of the data. Typical data of this type obtained by Lee and Srinivasan (1978) are illustrated in Figure 6.46.

A problem with the burst amplitude method is that changes in calibration can occur due to obscuration of windows etc. A technique which depends on ratio measurements rather than on absolute amplitude measurements is the so-called "visibility" technique of Farmer (1972) the principle of which is illustrated in Figure 6.47.

A droplet of small size compared with the width of the dark and light fringes generated at the beam-crossing point in a laser anemometry system will give a scattering signal fluctuating between zero intensity and the maximum amplitude (Figure 6.47). A larger droplet, on the other hand, will always lie partly within one of the light fringes and will always be scattering, though the scattered intensity will also be modulated at a frequency depending

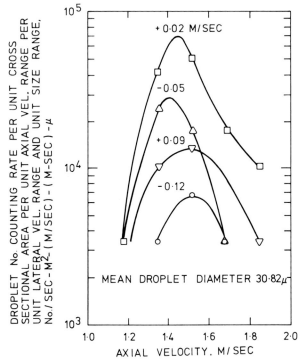

Figure 6.46: Droplet count rate related to lateral and axial velocity of the droplets (Lee and Srinivasan (1978)).

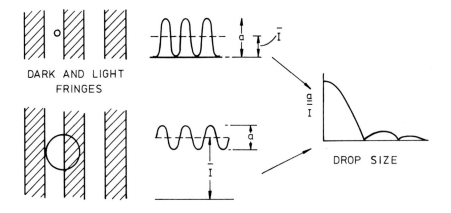

Figure 6.47: The Farmer (1974) visability technique for drop size measurement.

on the droplet velocity. The basis of the Farmer visibility technique is that the ratio of the modulation amplitude, a, to the average intensity $\bar{I}$ is a function of drop size as sketched in Figure 6.47. Thus, by measuring both the amplitude of the modulation and the mean intensity of the scattered light, an estimate of the drop size can be obtained. Applications of the visibility method are described by Birchenough and Mason (1976) and by Schmidt et al (1976). Due to the nature of the signal (see Figure 6.47) there is an ambiguity in the response for particle sizes and for particle diameters greater than 1.05 fringe spacings (corresponding to a/I of 0.15); this limits the method but many sprays can still be measured (eg Schmidt et al made measurements of drop size up to 90 μm).

A system for simultaneous drop size and velocity measurement which can be used for larger drops is described by Wigley (1977) and is illustrated in Figure 6.48. A droplet passing through the scattering volume (ie the intersection volume of the two laser beams) gives a back-scattered modulated signal from which the velocity can be deduced in the normal way. The forward scattering of the beams also occurs and the forward scatter signal shows three peaks from the leading edge, centre and trailing edge of the droplet respectively. By using a slit aperature, it is possible to avoid signals from all droplets which do not pass axially through the beam intersection point, thus avoiding ambiguity. Since the velocity is known from the modulated back-scatters signal, the crossing time between the first and third peaks multiplied by the velocity gives the droplet diameter. Though it is quite

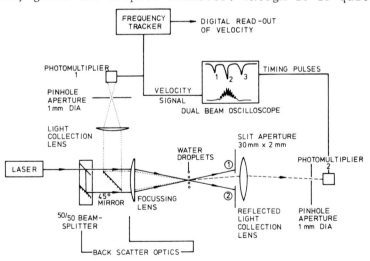

Figure 6.48: Technique for simultaneous drop velocity and drop size measurement (Wigley (1977)).

straight forward to measure the droplet size from photographic records of the received signals, it is more difficult, with this technique, to develop an automatic processing system.

Clearly, there is great scope for continued development of drop sizing methods linked to laser anemometry and much new light will be thrown on two phase flow phenomena by the use of these methods.

Miscellaneous Methods

The importance of droplet size measurement has led to the development of a variety of other methods which includes:

(1) Thermal droplet-size measurements. By allowing a stream of droplets to impinge on a heated surface, their rate of evaporation can be observed (in the Leidenfrost mode) and their size determined from the evaporation time. This technique has been reported by Leonchik et al (1970) and by Pimsner and Toma (1977). Van Paassen (1974) describes a device which measures droplet size by recording the response of a hot thermocouple following droplet impingement upon it. The technique was calibrated using artificially generated droplets and is claimed to give reproducible results.

(2) Ludewig (1972) describes a photoelectric method for measuring size distribution of dispersed drops in an immiscible binary liquid mixture. The probe collects the drops into a capillary and the length of liquid slugs in the capillary are determined photoelectrically.

(3) Bubble size from bubble velocity. Ben-Yosef et al (1975) describe a technique for the measurement of bubble size distribution, by measuring the bubble velocity distribution and deducing the size distribution from the relationship between rise velocity and bubble size.

(4) The needle-contact (conductance) method for local void fraction velocity and bubble size was described above. Drop size can be obtained from this method provided the contact hysteresis effect can be accounted for. For instance, Hoffer and Resnick (1975) used the method for drop size measurement in liquid/liquid systems.

The measurement of the size of fluid elements in dispersed flows is not absolute and, even for solid suspensions, there is variation between the various methods of measurement. The user has to select a technique best suited to his needs. Usually, some form of photographic technique can be used. In some circumstances, the magnesium-oxide coating technique can be used and the recent developments of the light scattering methods and laser anemometry related methods appear to hold considerable promise.

Further work needs to be done on drop size measurement using optical techniques, including laser methods.

Wall Shear-Stress

In single-and two-phase flows, it is customary to deduce the wall shear-stress through pressure gradient. However, the assumptions that need to be made, in deducing wall shear-stress, are considerable. One has to subtract from the total pressure gradient, a usually significant gravitational term and a usually only vaguely known accelerational term. In non-axisymmetrical channels, the wall shear-stress varies around the periphery of the channel and the same applies to non-symmetrical flows (such as horizontal two-phase flows), even in axisymmetrical channels. Direct measurements of both average and local wall shear-stress is thus advantageous, though there are considerable difficulties in applying the techniques developed in single-phase flows, to the case of two-phase flow. The three main types of technique employed are balancing methods, heat and mass transfer methods and impact probe methods. A general review of shear stress measurement in two-phase flow is given by Moeck (1969).

Balancing Methods

In these methods, the total force exerted by the fluid on the channel wall is measured by allowing the whole or part of the channel to "float" and to measure the force on it. This technique has been applied by Armand (1946) and Rose and Griffith (1964), who measured the shear force on the whole tube. In the work of Rose and Griffith, the force was determined using an electronic load-detection cell. The use of the balancing method, over relatively short elements of channel, has been described by Cravarolo et al (1964). The "τ"-meter they developed, is illustrated in Figure 6.49 and uses a freely suspended element of tube, contained in an outer annular space. Small gaps exist between the freely suspended part of the channel and the immediately upstream and downstream portions and a liquid purge passes through these gaps. Half-way up the suspended section, facing outwards towards the wall of the annular containing cell, there is a ring attached to the suspended part which splits the outer cell into two annular chambers. With a differential pressure of known value between these two chambers, it is possible to estimate the force required to maintain the element of tube in a freely suspended condition. This device has proved robust and has been used in a wide range of experiments and gives excellent accuracy. It can only operate where the fluid adjacent to the wall is likely to be liquid (eg annular flow). A similar device has been used by Stephens (1970) for shear stress measurements in an annulus and by Kirillov et al (1978).

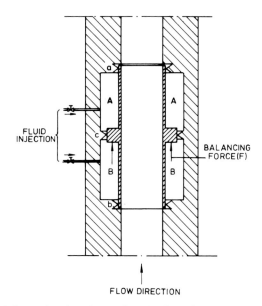

Figure 6.49: Schematic drawing of a device for the measurement of shear stress on the tube wall in two-phase flow (Cravarolo et al (1964)).

Heat and Mass Transfer Methods

The velocity profile immediately adjacent to the wall depends, for a given fluid, only on the wall shear-stress. Thus, Lavalee and Popovich (1974) have examined fluid flow close to the wall of a roughened channel, by means of a flash photolysis technique. Direct determination of the wall shear stress was claimed to be possible, from the data obtained.

Often, a localised and continuous measurement of shear stress is required and, for these purposes, the heat and mass transfer methods are potentially suitable. Shear stress measurements in a horizontal rectangular channel, using the hot film method, are reported by Shiralkar (1970). The principle of the method is illustrated in the sketch in Figure 6.50 and diagram of the probe used by Shiralkar et al is shown in Figure 6.51.

The probe heat flux $\emptyset$ and wall shear stress τ_o are related by the Leveque equation:

$$\frac{\emptyset}{(T_o - T_\infty)} = 0.538 k_L \left[\frac{\tau_o}{\mu_L \alpha_L x} \right]^{1/3} \quad (6.4)$$

where k_L is the liquid thermal conductivity, μ_L is the liquid viscosity and α_L the liquid thermal diffusivity

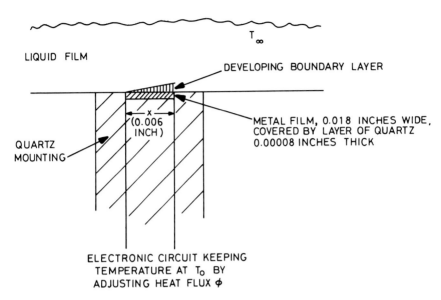

Figure 6.50: Hot film probe method for wall shear stress measurement.

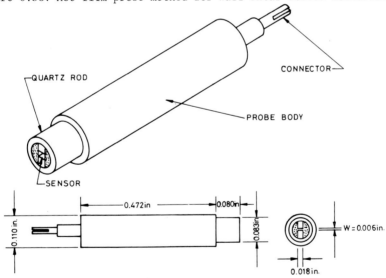

Figure 6.51: Hot film probe used for wall shear stress measurement in two phase flow by Shiralkar (1970).

(= $k_L/\rho_L C_{pL}$ where C_{pL} is the liquid specific heat). In attempting to apply hot film probe techniques at Harwell, a major difficulty was found to be the generation of gas bubbles on the probe tip by degassing of the liquid phase. For this reason, the probe should be operated at the lowest

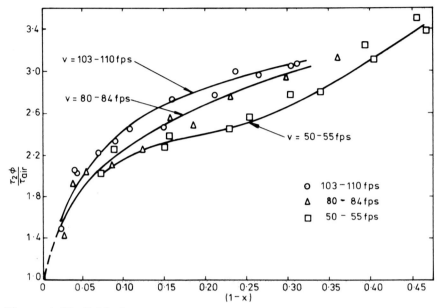

Figure 6.52: Wall shear-stress, related to the wall shear-stress of the air flow alone, for air-water parallel flow in a horizontal rectangular channel (Shiralkar (1970)).

possible temperature.

The most commonly used mass transfer technique is the electrochemical method; the basic principles of this method were described in Chapter 3 (see equations 3.7-3.9). The method of using this technique for wall shear stress measurement is indicated in Figure 6.53.

A very short section of surface is in the form of a nickel cathode, with a further (and very much larger) nickel anode situated also (typically) in the surface downstream of the measuring cathode. The mass transfer coefficient is estimated from the limiting current and bulk concentration (see equation 3.9) and this mass

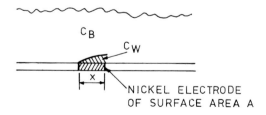

Figure 6.53: Electro-chemical method for wall shear stress measurement.

transfer coefficient is related to the shear stress by an equation analogous to Equation 6.4):

$$\bar{k}_c = 1.68 \, D_{AB} \left[\frac{\tau_o}{\mu_L D_{AB}} x \right]^{1/3} \qquad (6.5)$$

where D_{AB} is the diffusion coefficient of the ferricyanide ions.

The electrochemical method has been applied to wall shear stress measurement in two phase flow by a number of authors including Sutey and Knudsen (1969) and Kutateladze et al (1969) (1971) (1972) and Kutateladze (1969).

Impact Probe Methods

In aerodynamic experiments, it is common to use Preston (Preston (1954)) or Stanton tubes to measure wall shear-stress. These consist of tiny openings facing into the flow, adjacent to the wall. The impact pressure measured in these probes is characteristic of the velocity profile adjacent to the wall and, thus, of the shear stress. Shear stress measurements using Preston tubes have been reported for two-phase systems by Schraub et al (1969) and Davis (1969). The method appears to be satisfactory, providing the film is *always* thick compared with the size of the probe. Achenbach (1971) describes a skin-friction device, which may have some advantages over those Preston and Stanton tubes. It consists of a protrusion, of a few microns height, with slits on either side of it. The pressure drop across the protrusion gives an indication of the shear stress.

For measurements of wall shear-stress, it would seem that the balancing method, (particularly in the form described by Cravarolo et al (1964)), can give good and reproducible answers in those cases where it can be easily applied. For other systems, the diffusion-controlled electrolysis method has now been quite widely used and there have been some measurements on hot-film probes. The latter may suffer from susceptibility to bubble formation in the boundary layer, but can, of course, be used without the provision of special electrolyte liquid in the system. The use of all impact probes should be regarded with some suspicion in two-phase flows, particularly when the gas phase is likely to exist close to the wall.

Further work, including the measurement of wall shear-stress, would appear to be highly desirable in sorting out the still-unresolved problems in evaluating the various components of two-phase pressure drop.

Temperature Distribution

The temperature distribution in two-phase systems is of

considerable interest in studying heat transfer. For instance, measurements of temperature profiles in subcooled boiling can give valuable indications of the mixing processes involved, (see, for instance, Alekseev et al (1974)). However, when the two phases are not in thermodynamic equilibrium, the local temperatures measured by thermocouples are open to ambiguities. The various ways in which departures from equilibrium can occur were reviewed in Chapter 1.

The problems encountered in measurements include:

(1) *Impact temperature effect*. If a vapour is approaching a temperature sensor at velocity V and static temperature T_s, then, if the fluid is brought to rest on the sensor, the temperature recorded by the sensor is the total temperature T_t given by:

$$T_t - T_s = V^2/2C_p \qquad (6.6)$$

In many systems, the difference between T_t and T_s is significant. Normal sensors tend to measure a temperature which is intermediate between the two values. In order to measure the total temperature, Murdock and Fiock (1950) designed a steam temperature probe in which the sensor was mounted within a tube which had large holes drilled in it facing the oncoming fluid with small escape holes on the opposite side of the tube. This was found to give a good measurement of the total temperature.

(2) *Radiation from surrounding surfaces*. In many reactor safety experiments, fluid temperature measurements are required in channels, the walls of which are very hot. Radiation from these walls tends to be picked up by the sensor and it is often necessary to insert radiation shields in order to obtain the true temperature of the fluid. An example of such a system is that used by Cadek et al (1971); they used a thermocouple mounted in a series of concentric tubes and aspirated the vapour through these tubes in order to measure the local temperature.

(3) *Wetting of sensor*. If the steam temperature in a superheated vapour/droplet mixture is required, then the thermocouple is often wetted by the droplets and a temperature near the saturation temperature is recorded. The wetting effect could hopefully be eliminated by the use of suitable shields to divert droplets away from the sensor. However, due to recirculation effects around the shields, it is often practically impossible to guarantee that the sensor is not partially or intermittently wetted. Work on development of probes for such measurements is described, for instance, by Chen et al (1977) and a shielded probe design suggested

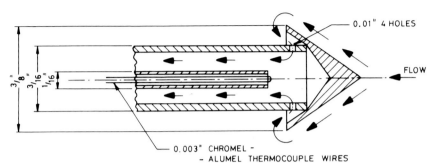

Figure 6.54: Probe for measurement of vapour superheat in the presence of liquid droplets (Chen (1977)).

by Chen et al is shown in Figure 6.54.

A technique which could also possibly be adapted to avoid the difficulty is described by Doe (1967) who constructed a thermcouple device for measuring the dew point of air/water vapour mixtures. The thermocouple was successively rapidly cooled and rapidly heated by the Peltier effect (passing a current in the reverse direction through the junction) and by ohmic heating (passing a current in the forward direction) respectively. In the interval between the cooling and heating pulses, the junction temperature could be recorded by displaying its emf on an oscilloscope. The effect of condensation on the junction was to cause a time varying signal of a recognisable type and, in Doe's experiment, the humidity could be calibrated against the size of the cooling pulse to produce this signal. To measure steam temperature, in the presence of water droplets, the thermocouple could be heated to prevent wetting (ie to above the Leidenfrost temperature) and then rapidly cooled to near the saturation temperature (in fact, condensation at the saturation temperature could be used to calibrate the instrument). On cessation of the cooling current, the thermocouple would then rapidly attain the steam temperature, which would be recorded before the heating current was reinstated. With a small enough thermocouple, the response time should be fast enough to record the steam temperature between droplet impingement. A droplet impingement would be recognised by a gross distortion of the output trace. The number of droplet impingements per unit time could also be recorded.

General design considerations for thermocouple systems are discussed in Chapter 3 and by Legendre (1974). Descriptions of alternative temperature-measurement systems, which might be more widely applied in two-phase systems, are given in Chapter 3 and by Brixy (1971), (thermal noise measurements), Riddle et al (1973), (platinum resistance thermometry), and Blair and Lander (1974) (infra-red temperature scanning). A bibliography on temperature measurement is given by Freeze and Parker (1972).

In a number of two-phase flows systems, local temperature profiles can be obtained using interferometric techniques. The use of a Mach-Zehnder interferometer for two-phase flow measurements is described by Adams and Meier (1970) and the use of laser holographic interferometry (see Chapter 3) is described by Mayinger and Panknin (1974). Raman spectroscopy can also be used for remote temperature measurement; the principles of this technique were described earlier in this chapter in the context of the measurement of concentration distribution. By determining the ratio of the Stokes and Antistokes emission intensities it is possible to determine local temperatures; an example of local temperature measurement by this means is the work of Nibler et al (1976). The technique shows considerable promise for application in two phase flow situations.

Further development is required on local instantaneous temperature measurements. For some applications, further development of the laser holographic interferometry technique would appear to be desirable and, for others, more development of microthermocouples and other fast-response temperature indicators. Temperature measurements in non-equilibrium vapour-liquid systems could possibly be facilitated using Peltier cooling of thermocouple junctions and this technique seems worthy of investigation.

Entrainment (Film Flow-Rate)

In annular flow, part of the liquid phase is often entrained as small droplets in the gas core (see Chapter 1 for general description). Determination of the split of the liquid flow between the film and core is of vital importance in understanding and calculating annular flow. The methods available for determining the extent of liquid entrainment include:

(1) *Entrained flow measurement*. Here, the droplet flux is measured by using a sampling or isokinetic probe. The total entrained flow is found by integration and the film flow is the difference between the entrained flow and the total liquid flow.

(2) *Film removal method*. Here, the film is removed (usually together with some of the gas phase) through slits or patterns of holes on the wall or, alternatively, through a channel wall section which is porous. The forward momentum of the droplets carries them beyond the extraction device so that only the film flow is measured.

(3) *Dye mixing methods*. By adding a dye (or other tracer) to the film at a known rate, allowing the material to mix with the film and determining the concentration in the film, the film flowrate can be determined by a mass balance. This technique is hampered by the fact that mixing within the film is often rather slow and

mass interchange is taking place between the film and the gas core.

(4) *From measurements of film thickness and pressure drop.* A large number of experiments have demonstrated that film flowrate can be calculated with reasonable accuracy from film thickness and pressure drop measurements via the so-called "triangular relationship" (see Hewitt and Hall-Taylor (1970)).

(5) *Photographic methods.* Mayinger et al (1976) have made measurements of droplet flux by analysing cine records taken by the axial view method. Though this technique is very useful in so far as information on dropsize and radial velocity can be gleaned simultaneously, it is too tedious to use widely.

The entrainment flow measurement method has been used by a number of workers. Wallis et al (1963) and Magiros and Dukler (1961) have used single sampling probes and assumed uniform distribution of entrainment flow over the whole cross-section. This is not valid (Gill et al (1963a)). However, even if the radial distribution of mass flow is determined, the outer limit of integration is difficult to specify, since the probe begins to make contact with interfacial waves in the liquid film.

In applying the film removal method, the film extraction can be done through a slot in the channel wall. (Collier and Hewitt (1960), Gill et al (1963), Huyghe and Minh (1965), Paleev and Filippovich (1966), Akagawa et al (1976) and Ueda et al (1978). Alternatively, a section of the channel wall can be made porous and the liquid film sucked smoothly out of this. From the point of view of extraction of wavy films, the porous-wall method has advantages over the slot method, in so far as the waves are gradually reduced in amplitude as they pass over the porous section and are thus eventually extracted. Using the slot method, there is a danger of spill-over when the large wave passes. A typical film extractor is illustrated in Figure 6.55.

Since the liquid film flowrate is fluctuating, due to the presence of waves, not all of the porous section is wetted by the film, the upper regions being dry. The gas phase is extracted through this upper region. The normal method of assessing the efficacy of extraction of the liquid film is to plot the liquid flow extracted against the gas extract rate. Often, there is a plateau which is taken to represent the film flowrate. At high mass flows and low qualities, some special pleading is often necessary! The various methods of film flow-rate measurement were reviewed by Simpson (1969) who concluded that the porous sinter method gave less consistent results than did the slot or wall scoop techniques. This may be so, but in the opinion of the present author, the porous sinter is far more likely to give the true film-flow (including the liquid carried in the waves).

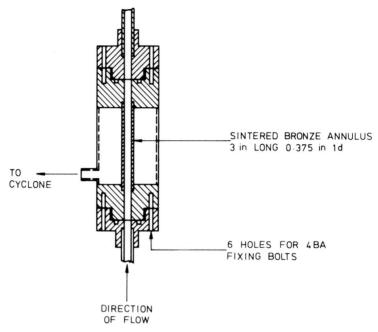

Figure 6.55: Porous wall film extraction device. Cousins et al (1965).

The application of the film suction method to single component fluids is more complex since the extracted two-phase mixture can flash and change composition. However, using a condenser/cooler, it is possible to convert all the outflowing mixture into the liquid form and, by carrying out a heat balance over the condenser, it is possible to determine the enthalpy, h_{TP}, of the outflowing stream. The principle is illustrated in Figure 6.56.

If the coolant inlet and outlet enthalpies (determined from temperature measurement) are h_{c2} and h_{c1}, the final enthalpy of the outlet stream h_o and the coolant and outlet stream flowrates are respectively W_c and W_o, the two-phase enthalpy is given by:

$$h_{TP} = h_o + \frac{W_c}{W_o}(h_{c2} - h_{c1}) \qquad (6.7)$$

Assuming that the mixture leaving the tube is saturated at the tube pressure p (with liquid enthalpy h_L and vapour enthalpy h_V), then the quality of the mixture leaving the channel is:

Entrainment (Film Flow-Rate)

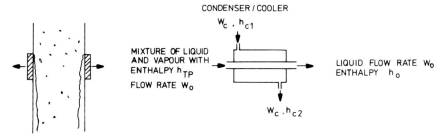

Figure 6.56: Film suction method for single component systems.

$$x = \frac{h_{TP} - h_L}{h_V - h_L} \qquad (6.8)$$

and the liquid and vapour rates flowing through the porous wall section are thus given by:

$$W_L = (1 - x) W_o \qquad (6.9)$$

$$W_G = x W_o \qquad (6.10)$$

The film flowrate is obtained by varying the extraction rate and plotting the liquid extraction rate versus the vapour extraction rate (in a similar manner to that used for air-water flows).
 In some situations, it is adequate to carry out a vapour/liquid separation of the extract stream and to measure the flows of vapour and liquid separately using flowmeters.
 Measurements of liquid film flow by the porous-wall suction method have been reviewed by Hewitt (1964), Hewitt and Hall-Taylor (1970) and Moeck (1969). The technique is particularly useful in the elucidation of mechanisms of the burn-out phenomenon (Hewitt et al (1965), Staniforth et al (1965), Keeys et al (1970), Jensen et al (1971), Moeck and Stachiewicz (1970), Singh et al (1969), Bennett et al (1969), Brown et al (1975), Nigmatulin et al (1977) and Subbotin et al (1978)). Typical data for film flow rate variation with power are shown in Figure 6.57. Note the approach to zero film flow rate as the burnout condition is approached.
 Difficulties are also encountered in applying the technique to down-flow systems (Webb et al (1970)), where the bottom of the porous section tends to flood.
 The film suction method can also be used to measure local film flowrate, For example, in a horizontal tube,

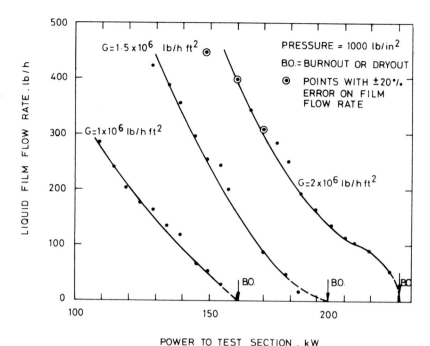

Figure 6.57: Liquid film flow rate data obtained by the film suction method for steam-water annular flow at 70 bar. Bennett et al (1969).

film flowrate in annular flow varies around the periphery of the tube due to the fact that liquid tends to flow towards the bottom of the tube under gravity. Typically, the liquid flow at the bottom can be in order of magnitude greater than that at the top. Measurements of film flow distribution using the film suction method are described by Butterworth (1969), Butterworth and Pulling (1973) (1974) and Maddock and Lacey (1974). The device used by Butterworth and Pulling is illustrated in Figure 6.58.

A segment of the film flow is separated using fins between which is sited a porous wall section through which the separated film segment is removed. The device is designed so that the position of measurement can be rotated around the circumference.

An alternative method for measurement of local film flowrate is described by Coney and Fisher (1976) and is illustrated in Figure 6.59. Potassium chloride solution is pumped at a constant rate into a sector of liquid film with which it mixes. The mixed concentration is measured using a miniature conductivity probe as shown in Figure 6.59: with this probe, the gap between the electrode is only 10μm and the probe is thus insensitive to variations in film thickness, its response depending only on the conductivity, and hence the concentration of potassium

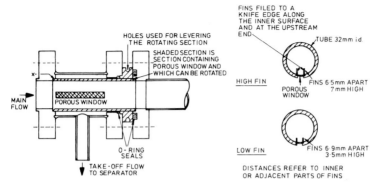

Figure 6.58: Local film flow rate measurement in horizontal annular flow. Butterworth and Pulling (1973).

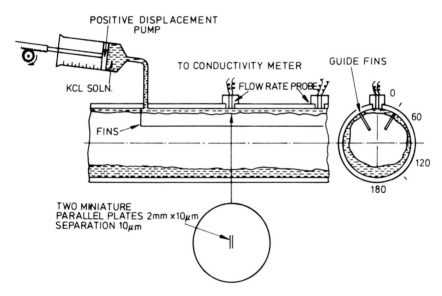

Figure 6.59: Salt dilution method for measurement of local film flow rate. Coney and Fisher (1976).

chloride in the liquid film. The effectiveness of the mixing was checked by having two successive probes.

To summarise, it can be stated that reproducible and accurate results can be obtained using film suction methods provided the flow is not too complex and the mass flux too high. Where the technique cannot be applied, it is worthwhile examining alternative possibilities such as dyestuff injection.

Contact Angle

In gas-liquid two-phase systems, where the two phases

are in contact with a solid boundary surface, an important parameter of the system can be the contact angle. Examples of such two-phase systems would be rivulent flow and drop-wise condensation. Measurements of contact angle can be made directly by observing the triple interface, in a direction parallel to the surface (Davies et al (1971)). A neater technique, based on reflection, has been used by Towell and Rothfield (1966). A lamp and an eye-piece are moved together, in a direction parallel to the surface on which the triple interface is situated. When the refleciton of the light is no longer seen, a line normal to the liquid surface, and passing through the triple interface, has been passed. It is then simple to evaluate the contact angle from the position of this line of normal reflection.

Contact angle is important in systems where rewetting or de-wetting is occurring. It is also relevant in governing the nucleate boiling phenomena on a surface. More information is required, and should possibly be sought, on the contact angle, under conditions of high pressure and temperature.

7 Third Order Parameters

Introduction

Parameters which describe or define the time variation of two phase flow quantities are defined, in the context of this book, as "third order parameters". Since there are often large scale temporial fluctuations of these quantities, a full understanding of two phase flows can only come if the nature of these fluctuations are fully understood. In this Chapter, the following topics are discussed:
(a) Film thickness - interfacial waves.
(b) Concentration of phases and components.
(c) Velocity fluctuations.
(d) Pressure fluctuations.
(e) Temperature fluctuations.
(f) Photographic observations of local phenomena.
(g) Fluctuations in wall shear stress.

A general review of statistical measurements in two phase flow is given by Delhaye and Jones (1975) and of the interpretation of average quantities by Delhaye and Achard (1977).

Perhaps the most useful technique for the analysis of fluctuating signals is that of correlation. Suppose that a fluctuating signal (pressure, say) p(t) was being recorded, together with another time-varying signal y(t) (which might be from another pressure transducer or, say, a film thickness probe). The "auto-correlation" function for the pressure signal is defined as:

$$R_{pp}(\tau) = \lim_{T \to \infty} \frac{1}{T} \int_0^T p(t)p(\tau+t)dt \qquad (7.1)$$

The "power spectral density" of the pressure signal, indicating its distribution within the frequency domain, f, is obtained by a Fourier transform of $R_{pp}(\tau)$ as follows:

$$G_{pp}(f) = 2 \int_{-\infty}^{\infty} R_{pp}(\tau) \exp(-i\, 2\pi f\tau)d\tau \qquad (7.2)$$

$$= 4 \int_0^\infty R_{pp}(\tau) \cos 2\pi f\tau d\tau$$

The "cross correlation function" between signals p(t) and y(t) is defined as follows:

$$R_{py}(\tau) = \lim_{T\to\infty} \frac{1}{T} \int_0^T p(t-\tau)y(t)dt \qquad (7.3)$$

Commercial units (eg the Hewlett Packard 3721A) are available for processing the signals to produce auto- and cross-correlation functions. Units are also available (eg the Hewlett Packard 3720A) for carrying out Fourier transforms of the signals. The type of inter-relationship between power spectral density functions and auto-correlation functions is illustrated in the sketches in Figure 7.1.

Note that when the power spectral density is a δ function (ie the signal has a single frequency) then the auto-correlation function is a cosine wave.

Film Thickness (interfacial waves)

In annular two-phase flow, the liquid film thickness varies continuously with time because of the presence of interfacial waves. These waves play a dominant role in the flow. Interfacial waves can be studied by photographic methods (Chung and Murgatroyd (1967), for example), or

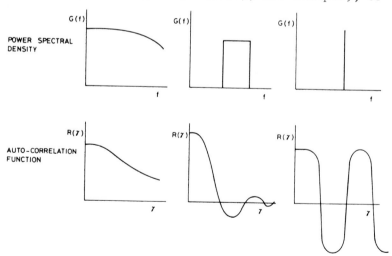

Figure 7.1: Typical power spectral density and corresponding auto-correlation functions.

their velocities can be measured by visual observation, using a mirror-scanner device (Hewitt and Lovegrove (1969)). The frequency of disturbance waves on the interface can be determined using needle-contact methods (Nishikawa et al (1967) (1968)) but the main technique that has been used, is that of instantaneous film thickness measurement. Four main techniques have been employed here: capacitance measurements, conductance measurements, light absorption methods and the fluorescence technique.

Measurements of interfacial structure, using capacitance techniques, are described by Tailby and Portalski (1960), for falling films on a vertical surface, The use of the technique for vapour-layer thickness measurement, in film boiling, is reported by Sheppard and Bradfield (1971). In both cases, a probe was inserted in the flow above the film surface.

The conductance probe technique for mean film thickness and film thickness probability-distribution measurements was reviewed in Chapter 6. By continuously recording the output from such probes, it is possible to get an indication of the interfacial characteristics, though the sensitivity problems tend to give an effective damping of the waves, in the output. This can be overcome, to some extent, by electronically adjusting the calibration curve, but the problem of the inability of these probes to make highly localised measurements still remains. However, the conductance-probe method is easy to apply and can be used in complex geometries. Building electronics for complex processing of the signals from such probes is relatively straightforward. The use of conductance probes, for the study of interfacial waves, has been discussed by Hewitt et al (1962), Hewitt (1969-70), Nishikawa et al (1967), Telles and Dukler (1970), Webb (1970), Zanelli and Hanratty (1971), Whalley et al (1973, Chu and Dukler (1974) (1975) and Brown et al (1975). Brambilla et al (1974) used conductance probes consisting of platinum wires stretched parallel across the tube bore and insulated except for the parts of the wires which were in one of the surface films. This seems to offer the possibility of disturbance of the film and the flush-mounted conductance probes are probably better. Telles and Dukler (1970), Webb (1970), Whalley et al (1973) and Brown et al (1975) all used conductance probes in determining the statistical properties of the interfacial waves. The output from the probe can be analysed to give a power spectral density of film-thickness fluctuation frequency. Examples of this kind of result are as follows:

(1) Figure 7.2 shows data obtained by Webb (1970) for downwards annular flow. The irregular behaviour of the interface in downwards annular-flow systems is well illustrated by the results. The power of the waves passes through a maximum, followed by a minimum, as the liquid flow rate increases.

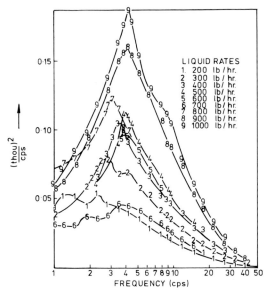

Figure 7.2: Power-spectra-density curves for film thickness in downwards co-current annular flow in a tube, (gas rate 150 lb/hr) Webb (1970)).

(2) Figure 7.3 shows data obtained by Whalley et al (1973) for upwards annular flow at an approximately constant mass velocity and varying quality. Note the increase in the wave power as the quality decreases and also the shift in peak frequency.

The velocity of the interfacial waves can be obtained by cross-correlation of the signals from two probes, placed at a known interval apart, in the direction of flow (Webb (1970), Telles and Dukler (1970) and Whalley et al (1973)). The cross-correlation function is the multiple of the two film-thickness readings at a known time interval and shows a peak at the characteristic velocity. Figure 7.4 shows the cross-correlation functions obtained for upwards air-water annular flow, with the probes mounted 0.5m apart.

Another technique, which has been quite widely used, is that of light absorption. A beam of light is passed through the liquid film and impinges, subsequently, on a filter detection device. Dyestuff is added to the liquid, so that the instantaneous intensity, of the received light, depends on the instantaneous absorption and, consequently, thickness of the liquid film. This technique presents geometrical difficulties, but these can be overcome by transmitting light through light-guides of various types (Hewitt and Lovegrove (1962)). The light absorption technique has been used by Charvonia (1962), Lilleleht and Hanratty (1961), Jones and Whitaker (1966) and Portalski and Clegg (1972),

Film Thickness (interfacial waves)   179

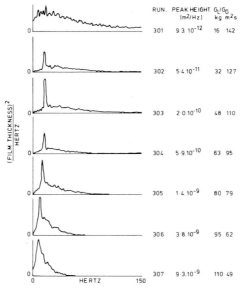

Figure 7.3: Power-spectral-density function for film thickness in upwards annular air-water flow, at a constant mass-velocity but varying quality (Whalley et al (1973)).

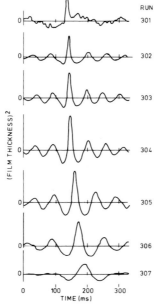

Figure 7.4: Cross-correlation function for film thickness, measured at two points 0.5m apart in upwards air-water annular two-phase flow (Whalley et al (1973)).

among others.

The technique apparently gives satisfactory results, providing the interface is not too disturbed. If the wave fronts are steep, then refraction of the incident beam can deflect the light, increasing its path length through the fluid or causing scattering away from the receiver. In an extreme case, Hewitt and Lovegrove (1962) observed that the refraction effects were completely dominant in co-current annular flow and entirely vitiated the technique. These interfacial refraction effects should always be checked carefully if this technique is used.

The refraction effects observed in the light absorption technique can be avoided by using the fluorescence technique. The latter was described in Chapter 6 (see Figure 6.10). It has the advantage over the light-absorption technique, that both the incident and received beams are in line and from the same side, ie it is possible to make film thickness measurements in the tube without the need to penetrate the film on the far wall with a light guide. The original method involved the use of a spectrometer for separating the received signal and is described in detail by Hewitt et al (1964), Hewitt and Nicholls (1969), Hewitt (1966) and Hewitt (1969-70). More recent developments of the technique have demonstrated the feasibility of using filtering systems rather than a spectrometer, and this gives a gross simplification in the instrumentation. Measurements using the fluorescence technique with special filter systems are described by Azzopardi and Lacey (1974) and Anderson and Hills (1974). The optical system used by Anderson and Hills is illustrated in Figure 7.5.

To summarise, it can be stated that the conductance probe method is probably the simplest, and the most convenient, technique for studying transient film thickness behaviour. The fluorescence method is nearest to ideal, and may now be easier to apply, after the development of suitable filtering systems. Other techniques, such as light absorption and capacitance measurement, should be treated with some reserve.

Concentration of Phases and Components

The local and average phase-concentration fluctuates in what are nominally steady two-phase flows, in a manner characteristic of the flow pattern. The problem of measuring rapidly time-varying void-fraction was discussed in the context of transient accident conditions in Chapter 4. The work of Jones and Zuber (1974) on the use of void-fraction fluctuations as a means of characterising flow patterns, was discussed in Chapter 6. In the same experiments, Jones and Zuber (1974) also measured the power spectral density of void-fraction fluctuations, using their X-ray absorption technique and some results obtained by them are illustrated in Figure 7.6. A characteristic peak in the void fluctua-

Concentration of Phases and Components 181

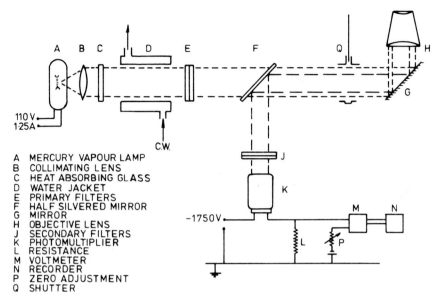

A MERCURY VAPOUR LAMP
B COLLIMATING LENS
C HEAT ABSORBING GLASS
D WATER JACKET
E PRIMARY FILTERS
F HALF SILVERED MIRROR
G MIRROR
H OBJECTIVE LENS
J SECONDARY FILTERS
K PHOTOMULTIPLIER
L RESISTANCE
M VOLTMETER
N RECORDER
P ZERO ADJUSTMENT
Q SHUTTER

Figure 7.5: Fluorescence method for the determination of local film thickness. Development of the method to use filter systems, rather than a spectrometer (cf Figure 6.10) (Anderson and Hills (1974)).

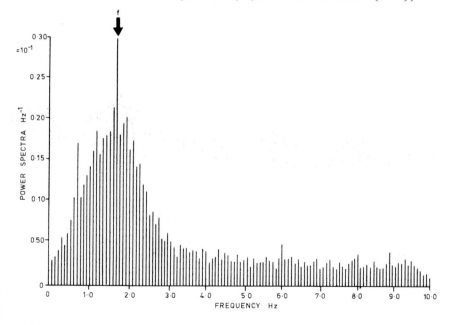

Figure 7.6: Power-spectral-density for void fraction in slug flow (Jones and Zuber (1974)).

ions is observed, in this case, at a frequency of 1-2 Hz. Continuous measurement of void fraction is also reported by Kolbel et al (1972), again using X-ray absorption methods. Other measurements of fluctuating void-fraction have been reported, using conductance needle-probes (Akagawa (1964), Akagawa and Sakaguchi (1966), Subbotin et al (1976), Subbotin et al (1977)), void impedance meters (Akesson (1968)), neutron absorption (Harms and Forrest (1971)), X-ray absorption (Smith (1975)), optical probes (Abuaf and Jones (1977)), capacitance probes (Rosehart et al (1975)) and beta absorption (Alekseev et al (1978)).

Measurements of the fluctuation of concentration of a component, transferrable from one phase to the other, are of interest in understanding the mechanisms of mass transfer. A system for measuring local concentration fluctuations is described by Torrest and Ranz (1969) (1970). These authors studied a liquid-liquid system and measured the local concentration fluctuations, using micro-conductance probes and optical probes.

In many two-phase mass transfer systems, material is absorbed into a liquid film. The absorbing material diffuses into the film and forms a layer adjacent to the film interface, which varies in thickness with time, due to the presence of interfacial waves. Knowledge of the thickness of the absorbed layer and of its time variation is of interest, particularly if simultaneous measurements of the film thickness are made. Measurements of this type are reported by Stainthorp and Wild (1967) and Schwanbom et al (1971). Stainthorp and Wild used a two-colour light absorption system in liquid-liquid mass transfer studies. A component was transferred from the liquid core into the film and gave absorption at one wavelength. The liquid film contained an untransferrable material, which absorbed at another wavelength. The latter absorption gave an indication of the liquid film thickness, and the former, the thickness of the mass transfer layer adjacent to the interface. Schwanbom et al (1971) used a fluorescence technique, to study the mass transfer behaviour of a gas-liquid system. Fluorescence was induced at two frequencies. The first frequency responded to an unreactive fluorescent dyestuff, mixed uniformly over the whole film; this signal gave an indication of the film thickness. The second frequency of emission corresponded to a region adjacent to the interface in which a fluorescent indicator, dissolved in the film, had been converted due to ammonia absorption in the film. The second signal gave an indication of the mass-transfer layer thickness. This method is illustrated in Figure 7.7 and might prove effective in studhing film mass transfer in systems more complex than that studied by Schwanbom et al.

Statistical analysis of local void fluctuations can give valuable information about the nature of the flow.

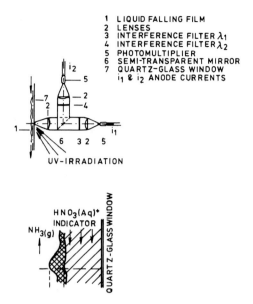

Figure 7.7: Simultaneous measurement of film thickness and mass-transfer layer (Schwanbom et al (1971)).

Similarly, concentration fluctuation measurements can give an indication of the mechanisms of component mass transfer. A number of techniques have been developed for both kinds of measurement. Further development of the X-ray void fluctuation device is indicated. This is consistent with the recommendation on flow patterns given above.

Velocity Fluctuations

Many of the techniques described in Chapter 3 above, for local velocity measurement, are readily applied to the determination of velocity fluctuations, resulting from turbulence and other features of thw two-phase flow. General studies of turbulent fluctuations, using heated sensors (hot wire, hot film, etc), are reported by Franktisak et al (1969), Gunkel et al (1971) and Wasserman and Grant (1973).

The hot-film anemometer can be used to measure the turbulent characteristics of the liquid phase in two-phase bubbly flow (Delhaye (1972)) in addition to its use in determining local void fraction.

Applications of hot film and hot wire sensors in the liquid layer of gas/liquid stratified flow are described by Jeffries et al (1969-70) and by Ueda and Tanaka (1975). The probe used by the latter authors is illustrated in Figure 7.8(a) and typical results for velocity fluctuations are shown in Figure 7.8(b); note how the profile of fluctuating velocity in the liquid layer deviates from

184   Measurement of Two Phase Flow Parameters

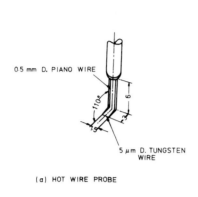
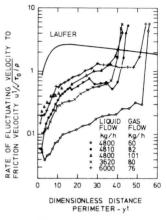

Figure 7.8: Measurement of velocity fluctuations in the liquid layer of a gas/liquid stratified flow (Ueda and Tanaka (1975)).

that calculated for single phase flow.

Electronic circuits for correcting hot-wire anemometer outputs for droplet impingements are described by Hetsroni et al (1969). The large pulses associated with droplet impingements on the anemometer wire are identified and separated and the turbulence intensity of the gas flowing over the wire between impingements can then be determined. Measurements using hot film anemometry, in horizontal separated flows, are reported by Jeffries (1969). Information on fluctuations in velocity can often be obtained by photographic methods. In fluid-fluid systems, tracer particles can be added to the liquid phase and their motion observed by photographic techniques. For instance, Goldsmith and Mason (1963) studied the liquid flow in the continuous phase, which was associated with the passage of a liquid bubble in liquid-liquid slug flow. Azzopardi and Lacey (1974) and Azzopardi (1977) used stereoscopic photography, to observe the velocity fluctuations occurring in liquid films, during the passage of waves. Cook and Clark (1971) describe an alternative method using tracer particles and photography which could be also used for this purpose. Currie and Rhodes (1971) describe a flash photolysis method for determination of velocity components in wavy films in annular flow.

The laser anemometry technique appears ideally suited

for measuring particle velocity fluctuations and, indeed, the application of the technique in single phase flow depends on the existence of minute particles in the flow, which scatter the light and which are assumed to follow the local fluid velocity. Widespread use of the laser anemometry technique, in the study of dispersed flows, is now taking place. Examples of work in this area are the studies of gas/solid flows by Raasch and Umhauer (1977) and Birchenough and Mason (1977), of gas/droplet flows by Lee and Srinivasan (1977) of gas/liquid bubble flows by Durst and Zare (1975) and of annular flow in triangular ducts by Ryzhkov and Khmara (1978). A detailed description of the laser anemometry technique was given in Chapter 6.

The laser anemometry technique appears ideally suited for measuring particle velocity fluctuations, provided the dispersed phase is not too concentrated. Other technqiues have application, also, in limited ranges (eg hot wire anemometry in highly-dispersed droplet flow).

Pressure Fluctuations

In most gas-liquid two-phase flow systems, the local pressure fluctuates with time, often by a large amount. Measurements of pressure fluctuations in two-phase flow have been reported, for instance, by Semenov (1959), Hubbard and Dukler (1966), Nishikawa et al (1969), Webb (1970), Akagawa et al (1971), Turner (1975), Dukler and Hubbard (1975), Chu and Dukler (1975) and Simpson et al (1977). Hubbard and Dukler and Simpson et al used statistical analysis of wall-pressure fluctuations to identify flow patterns in two-phase flow (see Chapter 6). Webb showed that the pressure fluctuations, in downward annular two-phase flow, were caused mainly by interaction of disturbance waves on the liquid film, with the outlet system. By smoothing the outlet, he reduced the fluctuations by an order of magnitude. The link between disturbance waves and pressure fluctuations was established by cross-correlation of the pressure and film thickness (conductance-probe) signals.

Kinoshita and Murasaki (1969) report measurements of fluctuating *dynamic* pressure at a Pitot probe in gas-liquid flow. By carrying out correlation analysis on the signals, it was possible to identify definite periodicity in the pressure fluctuation process.

Temperature Fluctuations

Even in what is nominally steady-state heat transfer, temperature fluctuations occur in two-phase flow systems, due to turbulence and the intermittency of phase content at a given location. The accurate measurement of local temperature fluctuations is usually achieved by the use of thermocouples made from small diameter wires (microthermocouples). A review of the applications of micro-thermocouples is given by Delhaye and Jones (1975). The micro-thermocouple throws considerable light on the mechanisms of nucleate boiling from the surface and studies of nucleate boiling using this technique, are reported by Semeria and Flamand (1967) (1970), Roemer (1970), Garg and Patten (1965), Franklin (1961), Delhaye et al (1972), Afgan et al (1973), Alekseev et al (1974) and Afgan (1976) (1978). Thermocouples down to 20 microns in diameter have been employed. Comparisons of the response of micro-thermocouples with that of miniature sheathed thermocouples were made by Siboul (1976). The micro-thermocouple used by Siboul is illustrated in Figure 7.9. Results from temperature fluctuation measurements, using micro-thermocouples, can be conveniently expressed in terms of a probability-density function, which might, typically, show two peaks corresponding to the respective phases, as illustrated in Figure 7.10 for an ideal case of saturated boiling. Often, however, the contributions of both phases are difficult to separate and an advanced use of the technique is described by Delhaye et al (1973b), who measured temperature fluctuations using an insulated thermocouple, which simultaneously served as a local void-probe on the conductivity principle

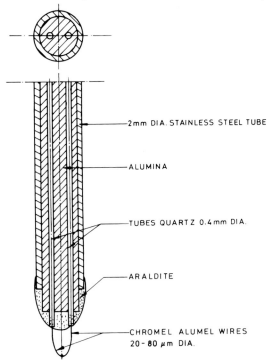

Figure 7.9: Micro-thermocouple construction used by Siboul (1976).

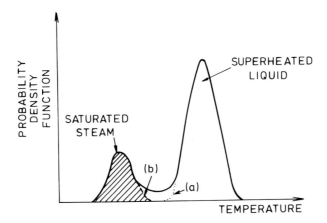

Figure 7.10: Typical probability-density function for local temperature, measured by a micro-thermocouple, obtained for saturated nucleate boiling.

(see Chapter 6). Using this device, the liquid-phase temperature distribution could be determined independently of the steam phase distribution, as illustrated in Figure 7.11. The lack of validity of earlier interpretations of micro-thermocouple data was demonstrated by this means.

The micro-thermocouple technique has been used by Barois and Huyghe (1969) in flashing flow (as applicable to desalination plants), and by Alekseev et al (1978) in flow boiling; fast-response thermocouples can be used for temperature measurement in a vapour containing liquid droplets. In the latter case, the method measures the super-heat of the vapour surrounding the droplets. The use of micro-thermocouples in liquid-liquid systems has been described by Delhaye (1970) and measurements on gas-

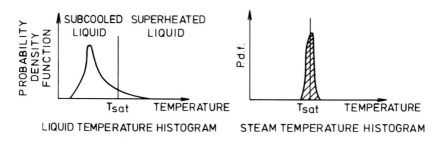

Figure 7.11: Separate probability-density functions for temperature of the liquid and vapour phases, respectively, in sub-cooled nucleate boiling; obtained by simultaneously detecting the local phases present, using a micro-thermocouple which acts as a needle void-probe (Delhaye et al (1973)).

solid flows are reported by D'Yachevskiy et al (1970).

The use of interferometric methods in finding temperature distribution was described in Chapter 6. Interferometric methods can be used for studying temperature fluctuations; an example of such work is that of Matekunas and Winter (1971) who applied the method to the study of nucleate boiling. Temperature fluctuation studies, using laser-holographic interferometry, are described by Mayinger and Panknin (1974).

During nucleate boiling, surface temperature fluctuations occur, associated with the formation and release of vapour bubbles. Moore and Mesler (1961) measured surface temperature fluctuations and found that, under some conditions, the temperature fell during the formation of the buble and then rose after bubble departure. This led them to postulate their "micro-layer evaporation" model for nucleate boiling. Further work on wall temperature fluctuation in nucleate boiling, has been reported by Hendricks and Sharp (1964), Rogers and Mesler (1964), Kovacs and Mesler (1964), Graham (1965), Marcus and Dropkin (1965), Morin (1965), Madsen (1965-66), Cooper and Lloyd (1966), Bonnet et al (1966), Deane and Rohsenow (1969), Fultz and Mesler (1970), Swanson and Bowman (1974) and Mesler (1975). Accurate measurement of wall temperature was difficult and several techniques have been developed. Cooper and Lloyd developed small surface-resistance thermometers, which were coated on the wall, using techniques developed in solid-state electronics. The response of these thermometers was insufficient to indicate that the wall temperature fell to the saturation temperature in the early part of the bubble release cycle. One would have expected the wall temperature to fall to saturation as the advancing front of the micro-layer passed over the temperature-sensing point. In further measurements, Fultz and Mesler used thin platinum films for resistance-thermometer studies, under growing bubbles, and this gave a faster response and, consequently, an indication of saturation temperature at one part of the cycle, as expected.

Measurements of temperature fluctuations, during dropwise condensation, were described by Takeyama and Shimizu (1974).

Information of considerable fundamental value has been obtained in boiling systems by the measurement of temperature fluctuations occurring in the fluid adjacent to the heat transfer surface (using micro-thermocouples) and in the solid surface from which the heat is being transferred (using thermocouples and surface-resistance thermometers). In systems where the structure's temperature is very different from the fluid temperature, radiation effects should be avoided. Information on temperature fluctuations can also be obtained using interferometric techniques coupled with cine-photography.

Further exploitation of the laser-holographic interferometry technique, for the study of fluctuating temperature fields, would appear to be worthwhile and promising. Also, measurement of local temperature in non-equilibrium systems should be pursued.

Photographic Observations of Local Phenomena

Usually, it is necessary to use high speed cine-photography to observe the time-varying nature of local phenomena in two-phase flow. A vast amount of work has been carried out using photographic techniques and it would be impossible, and inappropriate, to review this work in detail here. Work up to 1966 on the photography of two-phase flow was reviewed by Hewitt (1964) and Collier and Hewitt (1966) and developments in the field are described by Cooper et al (1963) and Arnold and Hewitt (1967). A useful review on photographic and other optical techniques is also given by Hsu et al (1969).

As was discussed in Chapter 6, complex refraction effects obscure photographs of two-phase flow used in delineating flow patterns. The same problem applies in observing local phenomena. A method of overcoming this is to use fibre-optic transmission. The application of this technique to studying droplet coalescence in liquid-liquid systems is described by Park and Blair (1975).

Typical of the many applications of high-spped photographic methods, in flows without evaporation or condensation, are the studies of Kordyban and Ranov (1963), Martin and Sims (1971) and Farukhi and Parker (1974). The study of the detailed motion of a dispersed phase, using photographic techniques, has been widely applied. For instance, Cumo et al (1971) studied the motion of small droplets in refrigerant-12 vapour in the post-burn-out region. Framing rates up to 40,000 per second were employed. The motion of individual bubbles in water streams has been studied by Baker and Chao (1965) and Sato (1974).

Numerous high-speed photographic studies have been made of two-phase evaporating flows and the problems involved are reviewed by Hewitt and Hall-Taylor (1970). Examples of such studies are those of Gunther (1951), Vohr (1963), Tippets (1964), Quinn and Swan (1964), Hosler (1965), Benenson and Stone (1965), Kirby et al (1965), Swan (1966), Bennett et al (1965-66), Mattson and Hammitt (1972), Mattson et al (1973), Bald et al (1977), van der Molen and Galjee (1977). High-speed photographic studies of condensation include, for instance, that of Keshock (1974), who studied flow condensation under normal-and zero-gravity conditions.

The photography of bubbling fluidised beds has been discussed by Lyall (1969). A method of following the motion of an individual particle in a fluidised bed, has been reported by Carlos and Richardson (1967). In these experiments, a liquid fluidised bed was used, in which

the liquid and the bulk of the particles in the bed, were
of the same refractive index. The bed was thus transparent
and when a dyed particle was introduced, its motion could
be observed and followed, using photographic techniques.

Photography of the liquid film, in gas-liquid flows.
has been used commonly in the study of interfacial waves.
By matching the refractive index of the flow in the liquid
(turpentine) and the channel wall (glass), Jacowitz and
Brodkey (1964) were able to photograph interfacial waves,
in a direction tangential to the circumference of the
tube, in which annular flow was taking place. The breakdown of the waves into droplets could be clearly observed.
A similar technique has been used by Truong Quang Minh and
Huyghe (1965) for the photographic determination of mean
film thickness.

An alternative view of the interface is obtained by
photographing in the axial direction. This technique is
discussed by Arnold and Hewitt (1967) and Hewitt and
Roberts (1969b) and the device for obtaining such photographs is illustrated in Figure 7.12. The technique has
also been used for viewing in curved tubes, where the
viewing direction is tangential to the curvature of the
tube (Azzopardi and Lacey (1974)).

More recently (Whalley et al (1977)) a stereoscopic
viewing system has been developed as sketched in Figure
7.13. Although the oblique angle view of the film has

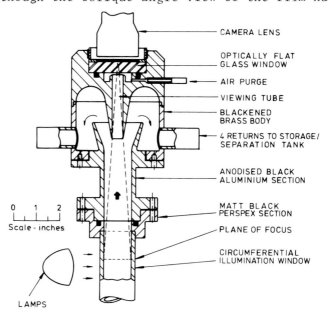

Figure 7.12: Device for axial-view photography of annular two-phase
flow (Hewitt and Roberts (1969)).

proved useful, the technique has limitations as a quantative tool since the depth over which the stereoscopic effect occurs is rather small.

In axial view photography of annular flow with droplet entrainment, it is not possible to identify the source of the droplets in the field of view. In annular flow, there is a critical liquid film flowrate below which the large, entrainment-creating, disturbance waves are not observed. If the flow is set at this critical value, so that naturally occurring waves are just absent, it is possible to inject a pulse of liquid around the periphery of the tube to create a single disturbance wave which then passes along the tube, generating entrained droplets. Using axial view photography, therefore, it is possible to study the droplet flux arising from a specific wave and this gives valuable information on entrainment characteristics. Experiments of this kind are described by Whalley et al (1977).

Other applications of the axial view photography method have included:
(1) Studies of adiabatic and evaporating refrigerant flows in a vertical tube (Mayinger and Langner (1977)).
(2) Studies of flooding in gas-liquid counter-current flow (Suzuki and Ueda (1977)).

Other studies of the onset of liquid entrainment in film flows, are reported by Woodmansee and Hanratty (1969) and by Yablonik and Khaimov (1972).

Photographic methods have been used in the observation of phase coalescence and breakup. Using two droplet generators and allowing the trajectory of the formed drop-

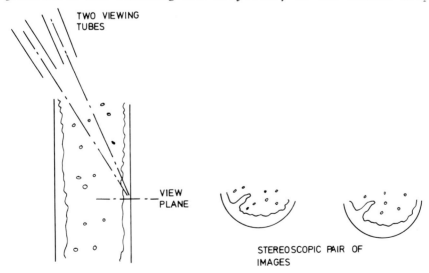

Figure 7.13: Stereoscopic axial view system (Whalley et al (1977)).

lets, to intersect, it is possible to induce the droplets to collide and to investigate their subsequent behaviour. Photographic studies of droplet coalescence have been reported by Park and Crosby (1965) and Ryley et al (1970). The combined droplet may, or may not, go into an oscillation mode which leads to its subsequent breakdown. Behar (1963) (1966) has reported on a stroboscopic method for bubble coalescence studies. Bubbles were released at regular frequency, from two adjacent orifices, and the various stages in the coalescence could be observed using a stroboscope, by suitably adjusting its phase with respect to the bubble emission. High-speed photographic studies of bubble coalescence are reported by Marucci et al (1968). When a bubble rises in a liquid, it drags with it, a weight of the continuous phase and such wakes have been investigated by Hendrix et al (1967). Swartz and Kessler (1970) report photographic studies of the breakup of individual droplets in liquid-liquid flows.

The study of static systems (eg pool boiling), can often give useful information on the interpretation of two-phase heat transfer and photographic techniques have been widely employed in studying such static systems. Interpretation of cine-films can give information such as bubble growth-rate and bubble release-rate. Typical of cine-photographic studies of pool boiling, are those of Gaertner (1963), Michiyoshi and Nakajima (1964), Torikai and Yamazaki (1965), Yu (1970), Tolubinskiy and Konstanchuk (1972), Subbotin et al (1974a) Cooper et al (1978, van Stralen and Zijl (1978), Oker and Merte (1978) and Hahne and Diesselhorst (1978). Yu studied boiling from patterns of artificial cavities and showed that there was no interference between cavities for distances, between cavities, greater than twice the bubble size at departure.

It is impossible to see inside the nucleation cavities on ordinary boiling surfaces and, in order to investigate the entry of the liquid phase into the cavity, (with the possible destruction of the cavity as a nucleation site), Kosky and Henwood (1968) photographed boiling taking place from a cavity consisting of the end of a heated capillary (acting as the cavity) for sub-cooled boiling. Similar experiments have been reported by Marto and Sowersby (1970).

In order to observe, directly, the micro-layer evaporation mechanism for the formation of bubbles, Torikai et al (1964) used a transparent heater to observe the bubble formation from below. They confirmed the formation of dry patches, due to micro-layer evaporation.

Studies of vapour-bubble growth, in superheated liquid pools, are reported by Abdelmessih (1968) and Beer (1971). Bubble growth in film pool-boiling was studied, using photographic means, by Bradfield (1966) and bubble-collapse phenomena were investigated by Green and Mesler (1970), using high-speed photographic methods, at framing rates up to 35,000 frames per second.

High-speed photography of static droplet systems is reported by Baumeister (1966), who studied the metastable Liedenfrost states of a drop on a hot surface, and by Adams and Pinder (1972), who photographed individual evaporating drops. Photographic studies of dropwise condensation are reported by Nijaguna and Abdelmessih (1974) and Reisbig (1974).

Photographic observations of two-phase heat transfer phenomena, can be greatly extended by taking advantage of the changes of refractive index, which take place as a result of density changes. This type of technique is particularly useful in studying thermal gradients, in both gaseous and liquid systems. Three distinct methods can be distinguished:

(1) *Interferometric methods:* These have already been discussed above, in the context of measurements of concentration and temperature distributions. Coupling the technique with high-speed photography, the local heat-transfer phenomena can readily be visualised. A further development of the technique, has been the use of holographic interferometry and this technique is reviewed by Mayinger and Panknin (1974). The interferometric methods depend on the fact that, if a light beam passes through a field in which the density is varying, the time of arrival at a point beyond the field will also vary, since the velocity of light through the field decreases with increasing refractive index. Interferometric measurements of the thickness of the micro-layer, under growing bubbles, are reported by Judd (1972).

(2) *Schlieren methods:* If there is a gradient of density, normal to the direction of the light rays through a flow field, then the rays will be deflected and this deflection can be observed and made to reveal the geometry of the density changes by using, say, the Toeplerschlieren technique. In this technique, light from a point source is focussed, using a large lens or mirror, onto a knife-edge and the subject to be observed is placed between the knife-edge and the focussing mirror or lens. Slight deflections of the light-rays, below the knife-edge, are then clearly observed by placing the eye, or the camera, on the far side of the knife-edge and focussing onto the mirror or lens. The schlieren technique can indicate temperature gradients in the boundary layer, adjacent to a surface. It is particularly useful in studying nucleate boiling and investigations using the technique, are reported by Behar and Semeria (1963a) (1963b), Jacobs and Shade (1969), Price and Bramall (1971) and Baranenko et al (1974). The technique can be further extended by using holographic-schlieren photography (Mayinger and Panknin (1974), Mayinger et al (1974)).

(3) *Shadow-graph methods:* If the gradient of the refractive index, normal to the light-rays passing through the subject, varies, then deflections of adjacent rays will differ and they will converge or diverge, giving increased illumination or decreased illumination on a screen, placed beyond the subject. This particular optical manifestation of density changes is, of course, the simplest to study experimentally, since all that is required is a point source of light (say a pin hole) and a ground-glass screen or photographic plate. The application of this technique to the study of boiling and degassing is reported by Behar and Semeria (1963a), (1963b).

To summarise, a great deal of work has been done using photographic techniques for studying two-phase flows. The studies have included both adiabatic and evaporating flow and a variety of special techniques have been developed. In the case of heated flows, simultaneous visualisation of the temperature fields can be obtained by using interferometric, schlieren and shadow-graph methods. The interpretation of photographic evidence from high-speed, highly dispersed flows is often very difficult.

Further evaluation of photographic techniques should continue. An example of this is the application of axial-view photography in annular flows. This technique could possibly be applied to high-speed transient blow-down flows.

Fluctuations in Wall Shear-Stress

Wall shear-stress fluctuations are of interest in the interpretation of pressure-drop data, in addition to giving an indication of the effect of turbulent fluctuations near the wall. Measurements of this type are reported by Lilleleht and Hanratty (1961), Kutateladze et al (1971), Kutateladze et al (1972), Kutateladze (1973) and Zanelli and Hanratty (1973). In the very extensive work of Kutateladze and co-workers, the diffusion-controlled electrolysis technique was employed, using small electrodes of platinum, typically 0.1mm thick and 1mm long. Results obtained by this technique are illustrated in Figure 7.14. Figure 7.14a shows the power spectral density of shear-stress fluctuations, as a function of frequency, for various flow conditions. Figure 7.14b shows the ratio of the mean fluctuation in shear stress to the mean shear stress and demonstrates a sharp peak in the slug-flow regime.

The technique of measuring instantaneous fluctuating values of shear stress, using the diffusion-controlled electrolysis method, is now well established and is throwing light on a number of phenomena in two-phase flow. The further application of the diffusion-controlled electrolysis technique would appear justified, particularly in the region of high mass-flows. For situations in which

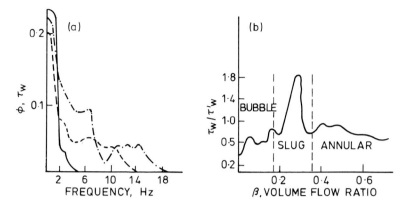

Figure 7.14: Shear stress power-spectral-density and relative fluctuation intensity in shear stress for two-phase flow, using the diffusion-controlled electrolysis technique (Kutateladse et al (1972)).

the use of the electrolyte solution is impossible, there may be further scope for the development of hot-film probes for local shear-stress fluctuation measurement.

8 Scaling of Two Phase Systems

Introduction

In the absence of prediction methods for two-phase flows, which are reliable enough to predict the behaviour of complex systems, the designer is often faced with the need to carry out direct, and often very expensive, experimental work on full-scale system simulations. The cost and time-scale of such experiments could be considerably reduced, if satisfactory and cheap modelling methods could be developed for two-phase systems. Scaling methods are, of course, extensively used in single-phase flow, (for example in aircraft development), but the situation in two-phase flow is highly unsatisfactory. The difficulty is that the parameters governing the two-phase flow are larger in number and less well understood than in single-phase flow.

One area where two-phase flow scaling has been successful, is that of modelling water systems using fluorinated hydrocarbons. Many boiling rigs have been constructed for fluorinated hydrocarbons and particular attention has been paid to the measurement of burn-out flux. The present Chapter will concentrate on this area but other examples of scaling methods in two phase flow are the work of Greenkorn et al (1967) who investigated two-phase flow in a Hele-Shaw model representing a porous medium, the work of Robertson and Sheridan (1970) who report air-water modelling of liquid steel processes and the work of Kubie and Gardner (1977) who used n-buty/acetate/water mixtures to model steam-water flows at very high pressures.

Application of Fluorinated Hydrocarbon Scaling

Water has a high latent heat and a high critical pressure. Thus, to carry out experiments on evaporation of water at high reduced pressure (ie in the range where nuclear and conventional boiling plant operate), expensive equipment is required for two reasons:
(1) Power supplies simulating the nuclear or fossil fuel source tend to be rather large. High current, low voltage power supplies are usually required and in the largest scale test rigs, power supplies of 10-20 MW are used with a consequentially high cost.

(2) The equipment must be designed for high operating pressure. This not only increases the cost of the experimental plant itself but it also increases the cost of visualisation and measurement systems used in the experiments.

In many fields of engineering science, modelling methods are used extensively to avoid the necessity for large scale prototypic testing and, naturally, it would be advantageous to find some method by which small scale, cheap experiments could be done to simulate the high pressure water systems. The obvious course is to use low latent heat, low critical pressure fluids as substitutes for water. Fortunately, a range of fluids exists which have suitable ranges of physical properties, which are relatively readily available and which are inflammable and relatively non-toxic. These fluids are the fluorinated hydrocarbon refrigerants (trade names "Freons" or "Arctons") and the use of these fluids in two-phase modelling studies was pioneered at the Atomic Energy Research Establishment, Winfrith, England in the early 1960's; since that time, a large number of rigs have been constructed throughout the world using these fluids. For refrigerant-12, for instance, it is possible to operate with the same vapour/liquid density ratios as water at 70 bar with a pressure of only 10.4 bar and with heat input levels of only about 6% of those for the water. The main refrigerants which have been used in these studies are listed in Table 8.1.

TABLE 8.1

Refrigerants commonly used in Heat Transfer Experiments

No	Formula	Molecular weight	Boiling point °C	Critical Pressure (bar)	Critical temperature (°C)
R11	CCl_3F	137.4	23.8	44.1	198
R12	CCl_2F_2	120.9	-29.8	41.2	112
R22	$CHClF_2$	86.5	-40.8	49.8	96
R113	$CCl_2F\text{-}CClF_2$	187.4	47.6	37	216
Water	H_2O	18.02	100	221.2	374

Information on the other refrigerants is available, for instance, from the Dupont Freon Product Information Bulletins.

For the assistance of the reader, approximate values of physical properties of the refrigerants listed in Table 8.1 are given in Tables 8.2-8.5 respectively. Table

TABLE 8.2

Refrigerant 11 Saturation Line Properties (Approximate)

Temp (K)	Pres- sure (bar)	LIQUID					VAPOUR					Latent heat (10^4 J/kg)	Surface tension (10^{-3} N/m) (dyne/cm)
		Dens- ity (kg/m^3)	Spec- ific heat (10^{-2} J/kg.K)	Thermal conduc- tivity (10^{-2} W/m.K)	Viscos- ity (10^{-4} kg/m.s)	Enthalpy (10^4 J/kg)*	Dens- ity (kg/m^3)	Spec- ific heat (J/kg.K)	Thermal conduc- tivity (10^{-3} W/m.K)	Viscos- ity (10^{-6} kg/m.s)	Enthalpy (10^5 J/kg)*		
260	0.221	1560	859	9.83	6.20	2.29	1.43	540	6.66	9.78	2.16	19.3	22.5
280	0.534	1520	872	9.27	4.94	4.00	3.24	555	7.25	10.4	2.26	18.6	19.9
300	1.13	1470	887	8.73	4.11	5.72	6.47	568	7.89	11.0	2.36	17.9	17.4
320	2.14	1420	902	8.18	3.52	7.46	11.8	581	8.60	11.5	2.46	17.2	15.0
340	3.74	1370	918	7.63	3.06	9.24	20.0	596	9.37	12.2	2.56	16.3	12.6
360	6.10	1310	936	7.07	2.70	11.1	32.2	614	10.2	13.0	2.65	15.4	10.3
380	9.45	1250	959	6.53	2.42	13.0	49.9	639	11.0	13.9	2.73	14.3	8.1
400	14.0	1190	999	6.04	2.20	14.9	74.9	676	12.0	14.9	2.80	13.1	5.9
420	20.0	1110	1080	5.62	2.02	17.0	111.0	743	13.3	16.1	2.87	11.7	3.9
440	27.7	1010	1260	5.20	1.74	19.4	166.0	895	15.5	18.7	2.90	9.64	2.0

* Based on 0 for the saturated liquid at 233K (-40°C)

TABLE 8.3
Refrigerant 12 Saturation Line Properties (Approximate)

Temp (K)	Pres- sure (bar)	LIQUID					VAPOUR					Latent heat (10^4 J/kg)	Surface tension (10^{-3} N/m) (dyne/cm)
		Dens- ity (kg/m^3)	Spec- ific heat (J/kg .K)	Thermal conduc- tivity (10^{-2} W/m.K)	Viscos- ity (10^{-4} kg/m.s)	Enthalpy (10^4 J/kg)*	Dens- ity (kg/m^3)	Spec- ific heat (J/kg .K)	Thermal conduc- tivity (10^{-3} W/m.K)	Viscos- ity (10^{-6} kg/m.s)	Enthalpy (10^5 J/ kg)*		
240	0.876	1500	887	9.28	3.69	0.603	5.54	561	7.02	10.3	1.73	16.7	16.4
260	1.96	1440	909	8.46	2.98	2.39	11.8	576	8.22	11.3	1.82	15.8	13.5
280	3.84	1370	937	7.68	2.42	4.24	22.3	599	9.41	12.3	1.91	14.8	10.8
300	6.85	1300	976	6.93	1.97	6.16	39.2	635	10.7	13.5	1.98	13.7	8.1
320	11.3	1220	1040	6.18	1.62	8.17	65.7	692	12.3	14.8	2.05	12.4	5.7
340	17.7	1130	1150	5.42	1.32	10.3	107.0	787	14.4	16.6	2.11	10.8	3.5
360	26.3	1010	1380	4.59	1.04	12.8	176.0	991	17.5	19.3	2.13	8.47	1.6

* Based on 0 for the saturated liquid at 233K (-40°C)

TABLE 8.4

Refrigerant 22 Saturation Line Properties (Approximate)

Temp (K)	Pres-sure (bar)	LIQUID					VAPOUR					Latent heat (10^4 J/kg)	Surface tension (10^{-3} N/m) (dyne/cm)
		Density (kg/m^3)	Specific heat (J/kg.K)	Thermal conductivity (10^{-2} W/m.K)	Viscosity (10^{-4} kg/m.s)	Enthalpy (10^4 J/kg)*	Density (kg/m^3)	Specific heat (J/kg.K)	Thermal conductivity (10^{-3} W/m.K)	Viscosity (10^{-6} kg/m.s)	Enthalpy (10^5 J/kg)*		
180	0.0368	1540	1080	14.6	6.08	39.7	0.214	519	5.81	8.17	6.62	26.5	27.0
200	0.166	1500	1090	13.5	4.74	41.8	0.872	552	6.81	8.96	6.71	25.3	23.7
220	0.548	1450	1100	12.5	3.89	44.0	2.66	585	7.73	9.73	6.81	24.1	20.4
240	1.43	1390	1110	11.4	3.31	46.2	6.56	626	8.59	10.5	6.91	22.9	17.1
260	3.17	1330	1140	10.4	2.89	48.5	13.8	686	9.46	11.3	7.00	21.5	13.9
280	6.19	1260	1190	9.43	2.56	50.8	26.2	771	10.5	12.1	7.07	19.9	10.7
300	11.0	1190	1260	8.43	2.32	53.3	46.8	881	11.9	13.0	7.12	18.0	7.7
320	18.0	1100	1370	7.32	2.13	55.9	80.2	1040	14.2	14.3	7.16	15.8	4.9
340	28.1	990	1640	5.91	1.88	58.7	134	1420	17.9	17.9	7.17	13.0	2.5

TABLE 8.5

Refrigerant 113 Saturation Line Properties (Approximate)

Temp (K)	Pres-sure (bar)	LIQUID Density (kg/m³)	Specific heat (J/kg.K)	Thermal conductivity (10⁻² W/m.K)	Viscosity (10⁻⁴ kg/m.s)	Enthalpy (10⁴ J/kg)*	VAPOUR Density (kg/m³)	Specific heat (J/kg.K)	Thermal conductivity (10⁻³ W/m.K)	Viscosity (10⁻⁶ kg/m.s)	Enthalpy (10⁵ J/kg)*	Latent heat (10⁴ J/kg)	Surface tension (10⁻³ N/m) (dyne/cm)
240	0.0225	1690	831	8.74	17.3	0.56	0.211	600	5.83	7.75	1.74	16.9	23.9
260	0.0749	1650	851	8.32	11.8	2.26	0.650	614	6.38	8.41	1.86	16.4	21.6
280	0.205	1610	875	7.89	8.54	3.96	1.67	632	6.99	9.06	1.99	15.9	19.3
300	0.480	1560	905	7.49	6.52	5.73	3.69	655	7.65	9.81	2.11	15.3	17.1
320	0.992	1510	939	7.09	5.15	7.61	7.23	684	8.36	10.6	2.23	14.7	14.9
340	1.86	1460	979	6.68	4.16	9.57	13.0	724	9.13	11.2	2.35	14.0	12.7
360	3.20	1410	1030	6.24	3.41	11.5	21.9	776	10.0	11.8	2.48	13.2	10.6
380	5.17	1350	1090	5.80	2.83	13.5	35.0	843	11.0	12.4	2.59	12.3	8.6
400	7.93	1290	1160	5.38	2.38	15.4	53.8	930	12.0	13.5	2.69	11.4	6.6
420	11.7	1220	1270	4.99	2.03	17.4	80.8	1050	13.1	15.0	2.78	10.4	4.8
440	16.6	1140	1430	4.62	1.75	19.7	122.	1240	14.5	17.3	2.87	9.04	3.0
460	22.9	1030	1700	4.20	1.47	22.3	194.	1590	16.6	21.1	2.92	6.95	1.5

* Based on 0 for the saturated liquid at 233K (-40°C)

TABLE 8.6
Water Saturation Line Properties

Temp (K)	Pres-sure (bar)	LIQUID					VAPOUR					Latent heat (10^5 J/kg)	Surface tension (10^{-3} N/m) (dyne/cm)
		Density (kg/m^3)	Specific heat (J/kg.K)	Thermal conductivity (10^{-2} W/m.K)	Viscosity (10^{-4} kg/m.s)	Enthalpy (10^4 J/kg)*	Density (kg/m^3)	Specific heat (J/kg.K)	Thermal conductivity (10^{-3} W/m.K)	Viscosity (10^{-6} kg/m.s)	Enthalpy (10^5 J/kg)*		
300	0.035	996	4178	61.2	8.53	11.3	0.025	1892	19.4	9.24	25.5	24.4	71.7
330	0.171	985	4184	65.2	4.91	23.8	0.113	1921	21.4	10.3	26.0	23.7	66.8
360	0.621	968	4206	67.5	3.28	36.4	0.378	1991	23.6	11.5	26.6	22.9	61.3
390	1.79	946	4240	68.5	2.39	49.0	1.02	2111	26.3	12.7	27.0	22.1	55.5
420	4.37	920	4298	68.6	1.85	61.9	2.35	2298	29.6	13.9	27.4	21.2	49.4
450	9.32	890	4390	67.9	1.51	74.9	4.80	2581	33.8	14.9	27.7	20.2	42.9
480	17.9	856	4532	66.1	1.30	88.3	8.98	2995	38.9	15.8	27.9	19.1	36.3
510	31.7	818	4741	63.1	1.15	102.3	15.8	3585	45.7	16.9	28.0	17.8	29.3
540	52.4	773	5081	59.1	1.03	116.9	26.5	4463	55.0	18.2	27.9	16.2	22.2
570	82.1	719	5704	54.5	0.92	132.7	43.7	5870	69.5	19.5	27.6	14.3	15.1
600	123.5	649	7032	49.2	0.82	150.0	72.1	8981	95.7	21.2	26.9	11.8	8.4
630	179.9	544	12881	40.7	0.66	173.4	132.5	21524	151.9	25.4	25.2	7.83	2.6

*Relative to 273K (0°C)

8.6 gives, for comparison, the appropriate figures for water. It should be stressed that physical property data are constantly being improved and that recent sources should always be consulted in finalising the values used in interpreting experimental results.

The main application of modelling using fluorinated hydrocarbons has been in studying the critical heat flux (or burnout) phenomenon in flow boiling. Examples of the extensive work which has been done in this area are:

(1) *Round tubes:* Data for critical heat flux in vertical upwards flow are reported by Stevens et al (1964), Staniforth et al (1965), Stevens et al (1965), Staniforth and Stevens (1965-66), Ilic (1974) and Merilo (1977). Studies of critical heat flux in horizontal tubes have been reported by Merilo (1977), in inclined tubes by Roumy (1974) and in serpentine tubes by Fisher and Yu (1975).
(2) *Annuli:* Studies of critical heat flux for refrigerant -12 evaporation in annuli are reported by Stevens et al (1968), Stevens and Macbeth (1968) and by Riegel (1971).
(3) *Rod bundles:* Data for critical heat flux in rod bundles are presented by Stevens and Wood (1966), Heron et al (1969), Stevens and Macbeth (1970), McPherson (1971), Stevens and Macbeth (1971) and Ilic (1974).
(4) *Critical heat flux under transient (eg blowdown) conditions:* This case is, of course of importance in modelling loss of coolant accidents in water reactors. Data on transient critical heat flux have been obtained by Leung (1976) and Henry and Leung (1977). In these experiments refrigerant-11 was used in preference to refrigerant-12 due to its higher boiling point and the possibility of operating the outlet at a lower pressure. Data on blowdown were also obtained by Belda et al (1974) using refrigerant-12.

Although the main emphasis has been on critical heat flux studies, the use of refrigerants as modelling fluids has been employed in a number of other studies of two phase flow parameters. These have included:

(1) *Void fraction:* Measurements of void fraction for refrigerant 113 are reported by Nabizadeh (1975).
(2) *Phase separation:* Studies of boiler steam drums using refrigerant-12 are reported by Thomas (1976) and studies of entrainment and separation in nuclear reactor vessels are reported by Viecenz and Mayinger (1977) who also used refrigerant-12.
(3) *Pressure drop:* Studies of pressure drop in evaporating refrigerant flows are reported by Friedel and Mayinger (1974) and by Trela (1974).
(4) *Flow patterns:* Studies of flow patterns in refrigerant -12 vapour/liquid flows are reported by Mayinger and Zetsman (1976).

The use of refrigerant modelling fluids has, indeed, proved a very powerful tool in studying two-phase phenomena. It is important to view this tool not only in the narrow sense of scaling to water but in the general sense of having convenient fluids with which to develop and test general models for two phase flow.

Development of Scaling Laws for Use with Fluorinated Hydrocarbon Scaling

Concurrently with the extensive experimental studies using refrigerants, there has been continuous development of methods by which the results could be scaled to those for water. Reviews of scaling methods are given by Anderson and Pejtersen (1969), Boure (1972), Cumo et al (1974) and Mayinger (1977). It is inappropriate in the context of the present book to do more than briefly survey the three main methods, namely the use of empirical scaling factor, the use of dimensionless numbers and the use of phenomenological modelling.

Empirical Scaling Factors for Critical Heat Flux

Perhaps the most widely used set of empirical scaling factors are those obtained in the early Winfrith work by Stevens and Kirby (1964) and Stevens and Wood (1966), relating data for refrigerant-12 to that for water at a pressure of 70 bars. The work of Barnett (1963) had indicated that, to obtain scaling between refrigerant and water, the ratios (ρ_L/ρ_G) and L/D) had to be fixed, where ρ_L and ρ_G are the liquid and vapour densities, L the channel length and D the channel diameter. Thus, to scale 70 bar water systems with refrigerant-12, the model system pressure needs to be 10.4 bar to obtain the same density ratio. In order to obtain modelling between the systems it is necessary to choose the correct values of the remaining <u>independent</u> variables - namely inlet subcooling ΔH, mass flux G, and channel diameter D. Partly on empirical grounds and partly on the basis of dimensional reasoning, the following scaling factors were derived:

$$F_G = \frac{\text{Mass flux for water}}{\text{Mass flux for refrigerant-12}} = 1.40 \quad (8.1)$$

$$F_{\Delta H} = \frac{\text{Subcooling for water}}{\text{Subcooling for refrigerant-12}} = 12.0 \quad (8.2)$$

$$F_D = \frac{\text{Channel equivalent diameter for water}}{\text{Channel equivalent diameter for refrigerant-12}} = 1 \quad (8.3)$$

Note that $F_D = 1$, implying that the same physical test

section size has to be used in the mode; this result is discussed in more detail by Stevens and Macbeth (1970) (1971). For given (ρ_L/ρ_G) and L/D) and with the ratios of independent variables specified by equations 8.1-8.3, the ratio of the heat flux for 70-bar water to that for 10.4 bar refrigerant-12 is given as $F_\emptyset$ = 16.8. Figure 8.1 shows comparisons of water and scaled refrigerant-12 data for rod cluster critical heat flux. As will be seen, excellent agreement is observed.

Dimensionless Groups for Scaling

The phase flows are highly complex and are exceedingly difficult to scale by the use of dimensionless groups. Complete similarity between systems is difficult to achieve, even in adiabatic flow, and in evaporating flows the dominant groups for scaling may change as the flow changes and develops along the channel. General discussions of scaling in two-phase systems are given by Chesters (1975) (1977) and by Ishii and Jones (1976). The use of dimensionless groups for scaling critical heat flux has been discussed by Barnett (1963) (1964), Ahmad (1973), Rinne (1975) and Nabizadeh (1977).

One of the most successful approaches to the problem

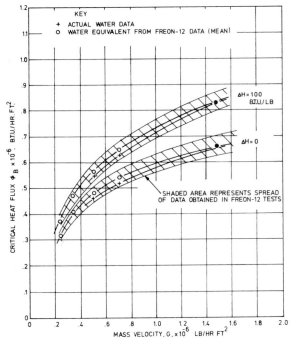

Figure 8.1: Comparison of data for critical heat flux for 70 bar pressure water evaporation in a 19 rod cluster with that obtained from scaled refrigerant-12 data (Stevens and Wood (1966)).

has been that described by Ahmad (1973). By classical dimensional analysis, Ahmad deduced that the critical heat flux (expressed in terms of the "boiling number" ($\emptyset/G\lambda$)) would depend on twelve dimensionless groups of which six were eliminated on inductive arguments. Three dimensionless groups are commonly matched in critical heat flux modelling:

Ratio of inlet sub-cooling to latent heat: $\Delta H/\lambda$

Density ratio: ρ_L/ρ_G

Length to diameter ratio: L/D

This left three dimensionless groups which were combined into a modelling parameter ψ_{CHF} obtained as follows:

$$\psi_{CHF} = \left[\frac{GD}{\mu_L}\right] \times \left[\frac{\mu_L}{\sigma D \rho_L}\right]^{2/3} \times \left[\frac{\mu_G}{\mu_L}\right]^{1/5} \qquad (8.4)$$

where μ_L and μ_G are the liquid and vapour viscosities, σ the surface tension (the other symbols were defined in the previous sub-section). An alternative to equation 8.4 is:

$$\psi_{CHF} = \left[\frac{GD}{\mu_L}\right] \times \left[\frac{\gamma^{\frac{1}{2}} \mu_L}{D \rho_L^{\frac{1}{2}}}\right]^{2/3} \times \left[\frac{\mu_G}{\mu_L}\right]^{1/5} \qquad (8.5)$$

where γ is $(\delta(\rho_G/\rho_L)/\delta p)$ where p is the pressure.

Equation 8.4 is the more fundamental version but equation 8.5 gives nearly identical results over wide ranges of pressure and temperature. The success of this modelling criterian may be judged from Figure 8.2 where data for CO_2, water and refrigerant-12 are compared (Hauptmann et al (1973)).

Figure 8.3 shows data obtained by Merilo (1977) for vertical and horizontal tubes. The Ahmad scaling method does not apply as well (as will be seen) to horizontal as it does to vertical pipes. This is probably representative of the different flow patterns obtained and illustrative of the fact that it is probably not possible to obtain satisfactory general scaling via dimensionless groups in evaporating flows, as was mentioned above. Dimensionless group scaling of pressure drop is discussed by Bruce (1972).

Predictive Methods

In recent years, there has been increasing interest in general predictive methods for two phase flows. An example

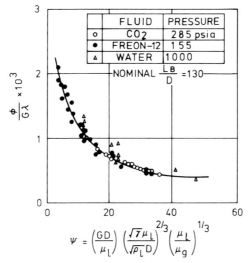

Figure 8.2: Application of Ahmad (1973) scaling method of water, CO_2 and Freon data (Hauptmann et al (1973)).

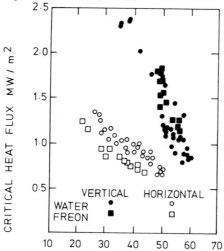

Figure 8.3: Comparison of vertical and horizontal critical heat flux data for a 1.26cm bore tube. Water pressure 6.9 MPa. Freon data modelled by Ahmad (1973) criteria (Merilo (1977)).

of these (described briefly in Chapter 1 and reviewed in more detail by Hewitt (1978)) is the modelling of annular flow. The prediction of the film flow rate along the channel is achieved by integrating the effects of liquid evaporation, droplet entrainment and droplet deposition.

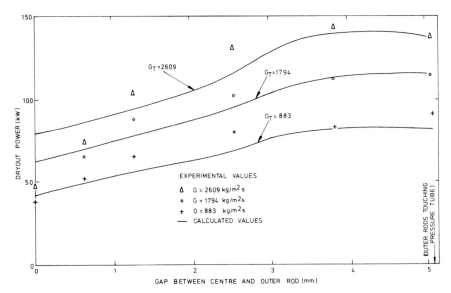

Figure 8.4: Comparison of experimental data obtained by Kinneir et al (1969) for critical heat flux in refrigerant-12 flow in a 7-rod cluster with values predicted by annular flow modelling (Whalley et al (1978)).

The existence of a large body of data for refrigerants has been of great assistance in testing the general validity of the methods developed. Good agreement is being found between the refrigerant data and the model predictions; this is exemplified by the results shown in Figure 8.4 for a 7-rod cluster. It is to be expected that experiments on refrigerants will continue to be of considerable assistance in the further development of basic modelling.

References

The page numbers on which the source is referred to are given in parenthesis at the end of the reference (eg [29,39] indicates that the source is referred to on pages 29 and 39).

Abdelmessih, A H (1968) Spherical bubble growth in a highly superheated liquid pool. *Int. Res. Symp. in cocurrent gas-liquid flow. Univ. Waterloo*, **2**. [192]

Abuaf, N and Jones, O C (1977) BNL instrumentation research program. *Paper presented NRC 5th Water Reactor Safety Res. Information Meeting, Gaithersburg, Maryland, USA, Nov. 1977*. [182]

Abuaf, N, Jones, O C, Zimmer, G A, Leonhard, W J and Saha, P (1978) BNL flashing experiments: test facility and measurement techniques *Paper presented at CNSI meeting on transient two-phase flow, Paris, June, 1978*. [49,126]

Achenback, E (1971) Influence of surface roughness on the flow through a staggered tube bank. *Warme-Und Stoffubertragung*, **4** (2) 120-126. [165]

Adachi, H (1974) High speed two-phase flow, *Jap. Atom. En. Soc. Japan*, **6** (6) 322-329. [72]

Adams, A E S and Pinder, K L (1972) Average heat transfer coefficient during the evaporation of a liquid drop. *Canadian J. Chem. Engng.* **50**, 707-713. [40,193]

Adams, J A and Meier, L D (1970) The effect of a submerged heater on the heat transfer near an evaporating surface. *Chem. Engng. Prog. Symp. Series* **66**, (102) 48-54. [168]

Adlam, J H and Dullforce, T A (1974) Spark-generated steam bubbles (experiment and theory) *UKAEA* CML-R134. [77]

Adler, P M (1977) Formation of an air-water two-phase flow. *A.I.Ch.E J.* **23**, 185. [122]

Adnams, D J, Salt, K J and Wintle, C A (1972) The development of instruments for the detection of dryout in uniform and non-uniform axially-heated rod clusters, *AEEW*-R574. [66]

Adorni, N, Alia, P, Cravarolo, L, Hassid, A and Pedrocchi, E (1963) An isokinetic sampling probe for phase and velocity distribution measurements in two-phase flow near the wall of the conduit. *CISE Milan*, Report No R-89. [118,127]

Adorni, N, Casagrande, I, Cravarolo, L, Hassid, A and Silvestri, M (1961b) Experimental data on two-phase adiabatic flow: liquid film thickness, phase and velocity distribution, pressure drops in vertical gas-liquid flow. *CISE Milan*, Report No R-35. [112,127]

Adorni, I N, Casagrande, I, Cravarolo, L, Hassid, A, Silvestri, M and Villani, S (1961a) Further experimental data on two-phase adiabatic flow: effects of the physical properties of the liquid phase-pressure oscillations - density measurements. *CISE Milan*.

Report No. R-41. [127]

Adrian, R J and Goldstein, R J (1971) Analysis of a laser-Doppler anemometer, *J. Physics E: Scientific Instruments*, **4**, 505-511 [135]

Aerojet Nuclear Company (1974a) Nuclear Technology Division annual report for period ending June 1974, *ANCR*-1177. [72,73]

Aerojet Nuclear Company (1974b) Development of instruments for two-phase measurements, *ANCR*-1181. [72,73,103,130]

Afgan, N H (1976) Boiling liquid superheat fluctuations. *In* "Two phase flow and heat transfer" *Nato Advanced Study Inst., Istanbul, Turkey, 16-27 Aug. 1976.* [186]

Afgan, N H (1978) Intermittent phenomena in pool boiling. *Proceedings of the 6th International Heat Transfer Conference, Toronto, Aug, 1978.* **6**, 407. [186]

Afgan, N H, Pislar, V, Javanovic, L J and Stefanovic, M (1973) An approach to the analysis of temperature fluctuations in two-phase flow, *Int. J. Heat Mass Transfer*, **16**, 187-194. [186]

Ager-Hanssen, H and Døderlein, J M (1958) A method of measuring steam voids in boiling water reactors: *2nd Geneva Conf. Peaceful Uses Atom. Energy. Geneva,* Paper 582, **11**, 463-465. [54]

Agostini, G, Era, A and Premoli, A (1969) Density measurements of steam-water mixtures flowing in a tubular channel under adiabatic and heated conditions, CISE Milan, Report No *CISE*-R-291. [61,62]

Agostini, V and Premoli, A (1971) Valvola di intercettazione rapida per imprego a aqua-vapore, *Energia Nucleare*, **18**. [61]

Ahmad, S Y (1973) Fluid to fluid modelling of critical heat flux: a compensated distortion model, *Int. J. Heat Mass Transfer*, **16**, 641-661. [205,206,207]

Akagawa, K (1964) Fluctuation of void ratio in two-phase flow, *Bull. JSME*, **7** (25), 122-128. [182]

Akagawa, K, Hamaguchi, K and Sakaguchi, T (1971) Studies on the fluctuation of pressure drop in two-phase slug flow, *Bull. JSME*, **14**, 447-467. [182,185]

Akagawa, K and Sakaguchi, T (1966) Fluctuation of void ratio in two-phase flow (2nd and 3rd Report), *Bull. JSME*, **9**, (35), 104-120 [182]

Akagawa, K, Sakaguchi, T and Ishida, T (1976) Creation and deposition of entrained droplets in swirling annular-mist two-phase flows. *Two-phase flow cavitation in power generation systems Conf. Grenoble, 30 March-2 April. 1976* 281-291. [169]

Akesson, H (1968) Local void measurements in oscillation two-phase flow. *Int. Two-Phase Flow Meeting, Oslo,* Paper B4. [92,182]

Alad'yev, I T, Gavrilova, N D and Dodonov, L D (1971) Effective electrical conductivity of two-phase gas-liquid metal flows. *Heat Transfer Soviet Research*, **3**, (4), 21-27. [63]

Alekseev, G V, Ibragimov, M Kh, Kulikov, B I, Sabelev, G I and Tarin, A A (1974) Temperature field in a flow of water in a circular pipe with surface boiling. *High Temp.* **11**, (6), 1103-1107 [166]

Alekseev, G V, Ibragimov, M G, Nevstrueva, E I, Ryabtsev, V A, Sabelev, G I and Tyutyaev, V V (1978) Some spectral characteristics of two-phase non-equilibrium flows. *Proceedings of the 6th International Heat Transfer Conference, Toronto, August, 1978,* **1**, 333. [182]

Alekseev, A V, Kazanskii, A M and Minailenko, A E (1973) Using a venturi tube to measure the concentration of the liquid phase in the case of flow of a two-component mixture. *Thermal Engng.* **20** (8), 86-89. [96,187]
Alexander, L G and Coldren, C L (1951) Droplet transfer from suspending air to duct wall. *Indust. Engng. Chem.* **45**, 1325-1331. [117,150]
Alforque, A, Sawada, Y and Mishihara, H (1977) Measure of acoustic velocity and attenuation in an air-water two-phase medium, *J. Nucl. Sci. Technol* **14**, No 2, 82-84. [71]
Alia, P, Cravarolo, L, Hassid, A and Pedrocchi, E (1968) Phase and velocity distribution in two-phase adiabatic annular dispersed flow. *Euratom* Report No. EUR-3759E. [118,127]
Alper, E (1976) Physical and chemical methods of measuring interfacial area in two-phase flow systems. *In* "Two phase flow and heat transfer", *Nato Advanced Study Institute, Istanbul, Turkey, Aug. 1976.* [152]
Alves, G E (1970) Cocurrent liquid-gas pipeline contactors. *Chem. Engng. Prog.* **66** (7), 60-67. [41]
Amenitskii, A N, Rinkevichyos, B S and Fabrikant, V A (1969) Laser measurement of the velocity distribution in a film liquid. *High Temp.* **7**, 974-975. [138]
Andeen, G B and Griffith, H P (1968) Momentum flux in two-phase flow, *J. Heat Trans.* **90**, 211-222. [87]
Andeen, G B and Kern, F R (1965) Two-phase momentum flux. 2. The heat pipe. *Progr. Report MIT DSR Project* No. 4547. [
Anderson, J D, Bollinger, R E and Lamb, D E (1964) Gas-phase-controlled mass transfer in two-phase annular horizontal flow. *A.I.Ch.E. Journal,* **10**, 640-645. [43]
Anderson, W K and Eno, B E (1960) Calculation of attenuation factors and percent signal change for X-ray void detection method in the limiting case of low quality mixed phase water systems, *KAPL-M-WKA-15.* [51]
Anderson, T T and Grate, T A (1969) Acoustic boiling detection in reactor vessels. *IEEE Trans. Nucl. Sci.* **16**, (1). [76]
Anderson, G H and Hills, P D (1974) Two-phase annular flow in tube bends. Symposium on multi-phase flow systems, Strathclyde, *I. Chem. E. Symp. Ser.* No. 38, Paper J1. [115,180,181]
Anderson, T T, Mulcahey, T P and Hsu, C (1970) Survey and status report on application of acoustic-boiling-detection techniques to liquid-metal-cooled reactors. *ANL-7409.*
Anderson, R and Pejtersen, V S (1969) A critical literature study on scaling laws for heat transfer and burn-out in two-phase flow with special reference to boiling-water reactors. *Danish AEC Riso* Report 207. [204]
Anderson, R J and Russel, T W F (1968) Film formation in two-phase annular flow. *AIChE Annual Meeting, Tampa, May, 1968.* [150]
Anderson, R J and Russel, T W F (1970) Circumferential variation of interchange in horizontal annular two-phase flow. *Industr. Engng. Chem. Fundam.* **9** (3), 340-344. [76,150]
Ando, Y, Naito, N, Tanabe, A et al (1975) Void detection in BWR by noise analysis, *J. Nucl. Sci. Tech.* **12**, No 9, 597-599. [64,141]

Andreoni, D and Courtaud, M (1972) Study of heat transfer during the reflooding of a single-rod test section. *Crest Specialist Meeting on Emergency Core Cooling for Light Water Reactors, Munich*, MRR 115, **1**. [82]

Andreussi, P and Zanelli, S (1976) Liquid phase mass transfer in annular two-phase flow. *Ing. Chim. Ital.* **12**, 132. [147]

Aoki, S, Inoue, A and Kozawa, Y (1974) Transient boiling crisis experiments. *Proc. JUICE Meeting Japan*. Report No 6. [79,80]

Appelt, K D and Kadlec, J (1975) Investigation of the resonance phenomena in the PS containment of the Marviken reactor during blow-down. Trans. *ANS*, **20**, 503. [72]

Arave, A E (1970a) An ultrasonic void fraction detector using compressional stress waves in a wire helix. *Idaho Nuclear Corporation*, Report No IN1441. [62]

Arave, A E (1972b) A conductivity-sensitive liquid level detector for reactor environment, *Aerojet Nuclear Company*, ANCR-1046. [103]

Arave, A E, Colson, J B and Fincke, J R (1977) Full flow drag screen. *US Nuclear Regulatory Commission, Proceedings of Meeting of Review Group on two-phase flow instrumentation, January 1977*, NUREG-0375 (Paper No I.5). [101]

Arave, A, Fickas, E and Shurtliff, W (1978) Ultrasonic density detector for in-core dynamic measurement. *Proceedings of the 24th International Instrumentation Symposium, Albuquerque, USA, May 1978*. [98]

Arave, A E and Goodrich, L D (1974) Drag-disk, turbine steam-water two-phase flow tests, *Nuclear Technology Division Annual Progress Report for period ending June 30, 1974*, ANCR-1177, 252-255. [98]

Ardon, K H and Duffey, R B (1977) A two-fluid continuum model for acoustic wave propagation in a non-equilibrium liquid-vapour flow. *CEGB Report No CEGB RD/B/N3916*. [71]

Aritomi, M, Aoki, S and Inoue, A (1977) Instabilities in parallel channel of forced-convection boiling of flow system (II) experimental results. *J. Nucl. Sci. Technol.* **14** (2), 14-22. [92]

Armand, A A (1946) The resistance during the movement of a two-phase system in horizontal pipes. *Izv. Vsesoyuznogo Teplotekhnicheskogo Instituta*, **1**, 16-23 (Translated UKAEA). [161]

Arnold, C R and Hewitt, G F (1967) Further developments in the photography of two-phase gas-liquid flow. *AERE*-R5318. [105,189,190]

Arrison, N L, Hancox, W T, Sulatisky, M T et al (1977) Blowdown of a recirculating loop with heat addition, *Conf. on Heat and Fluid Flow in Water Reactor Safety, Manchester, Sep. 1977*, proceedings 77-82 [72]

Arrowsmith, A and Foster, P J (1974) Hot-wire anemometry in miniature axisymmetric flows, *Journal of Physics E: Scientific Instruments*, **7**, 371-373. [128]

Artt, D W and Brown, A (1971) The simultaneous measurement of velocity and temperature, *Journal of Physics E: Scientific Instruments*, **4**, 72-74. [128]

Avidor, J M (1974) Novel instantaneous laser-doppler velocimeter. *Applied Optics*, **13** (2), 280-285. [138]

Azzopardi, B J (1977) Interaction between a falling liquid film and a gas stream, *Ph.D Thesis, Univ. Exeter*. [131,184]

Azzopardi, B J (1977) Measurement of drop sizes, *AERE*-R8667. [152]

Azzopardi, B J (1978) Consideration of the fluorescence film thickness technique, *Oxford University Engineering Laboratory*, Report 1229/78. [115]

Azzopardi, B J, Freeman, G and Whalley, P B (1978) Drop sizes in annular two-phase flow, *AERE*-R9074. [155,156]

Azzopardi, B J and Lacey, P M C (1974) Determining the structure of complex film flow. *Paper presented at European Two-Phase Flow Group Meeting, Harwell, June 1974*, Paper No A3. [131,180,184,190]

Azzopardi, B J, Serizawa, A and King, D J (1977) The behaviour and effect of bubbles within pressure tapping lines and orifices, *AERE*-R8792. [25]

Babukha, G L and Sergeyev, G I (1970) The effect of polydispersion of solid particles in a gas flow on the external heat transfer rate. *Heat Transfer Soviet Research*, **2** (1). [38]

Bailey, N A (1977) Dryout and post dryout heat transfer at low flow in a single tube test section, *AEEW*-R1068. [66]

Bailey, R V, Zmola, P C and Taylor, F M (1955) Transport of gases through liquid-gas mixtures, *ORNL*-CF-55-12-118. [49,50]

Bailey, S J (1974) Single ultrasonic beam measures liquid flow, *Control Engineering*, **21** (12), 53-54.

Baker, D J (1966) A technique for the precise measurement of small fluid-velocities. *J. Fluid Mech*. **26**, pt 3, 573-575. [132]

Baker, J L L (1965) Flow regime transitions at elevated pressures in vertical two-phase flow. *Argonne Nat. Lab.* Report ANL-7093. [106]

Baker, J L L and Chao, B T (1965) An experimental investigation of air-bubble motion in a turbulent water-stream. *AIChE Journal*, **11** (2), 268-273. [189]

Bakum, B I, Novikov, L V and Yakhlakov Yu, V (1968) Heat flux converter. *Measurement technqiues*, (3), 306-307. [38]

Bald, W B, Zust, H and Wooster, W G (1977) An apparatus for visualizing continuous flow boiling in liquid helium I, *Cryogenics*, **17** (1), 33-42. [189]

Banerjee, S (1978) A review of mass velocity measurements in two-phase flows, *Lecture at the Fluid Dynamics Institute, Short Course on Two-Phase Flow Measurements, Dartmouth College, New Hampshire, August, 1978*. [92,98]

Banerjee, S, Chan, A M C, Ramanathan, N and Yuen, P S L (1978a) Fast neutron scattering and attenuation technique for measurement of void fractions and phase distribution in transient flow boiling. *Proceedings of the 6th International Heat Transfer Conference, Toronto, August, 1978*, 351. [54,55]

Banerjee, S, Heidrick, T R and Rhodes, E (1978b) Development and calibration of instruments for measurements in transient two-phase flow. *Paper presented at CNSI Specialist Meeting on Transient Two-Phase Flow, Paris, June, 1978b*. [55,56]

Banerjee, S, Heidrick, T R, Saltvold, J R and Flemons, R S (1978) Measurement of void fraction and mass velocity in transient two-phase flow *In* "Transient two-phase flow" (Ed. S Banerjee and K R Weaver, *AECL, Toronto*, BP789-834. [98,128]

Bankoff, S G (1959) A variable density single-fluid model for two-phase flow with particular reference to steam-water flow. *ASME*

Paper 59-HT7. [9]
Bankoff, S G (1978) Vapour explosions: A critical review. *Proceedings of the 6th International Heat Transfer Conference, Toronto, August, 1978,* **6**, 355. [78]
Barenenko, V I, Chichkan, L A and Nikolev, G F (1974) Optical investigation of heat transfer mechanism with boiling. *5th International Heat Transfer Conf. Sci. Council, Japan,* **4**. [193]
Barnett, P G (1964) An experimental investigation to determine the scaling laws of forced convection boiling heat transfer. Part 1: The preliminary examination using burnout data for water and arcton 12. *AEEW-R363.* [205]
Barnett, P G (1963) The scaling of forced convection boiling heat transfer. *AEEW-R134.* [204,205]
Baroczy, C J (1976) A systematic correlation for two-phase pressure drop. *Chem. Eng. Prog. Symp. Ser.* **62** (44), 232. [11]
Barois, G and Huyghe, J (1969) Experimental studies of rising adiabatic flashing flow in a vertical channel of constant cross-section. *European Two-Phase Flow Group Meeting, Karlsruhe, June, 1969,* Paper 1.4. [187]
Barrett, P R (1974) Systematic errors in the discrete time-interval transmission method for the estimation of void statistics in boiling channels. *Nucl. Engng. and Design.* **30** (3), 316-327. [54]
Barschdorff, D (1977) Infra-red absorption method. *Report from the Univ. Karlsruhe, 1977.* [64]
Barschdorff, D, Class, G, Loffel, R and Reimann, J (1978) Mass flow measuring techniques in transient two phase flow *In* "Transient two-phase flow" (Ed. S Banerjee and K R Weaver), 835-867, AECL, Toronto. [101]
Bartolemei, G G and Chanturiya, V M (1967) Experimental study of the true void fraction when boiling subcooled water in vertical tubes. *Thermal Engng,* **14** (2), 123-128. [49]
Bartolomei, G G, Vinokur Ya, G, Kolokol'tsev, V A and Petukhov, V L (1960) Heat engineering and hydrodynamics. 2. utilization of gamma radiation in the study of the bubbling process. *USAEC Report No AEC-tr-4206,* 8-11. [49]
Baumeister, K J (1966) Metastable Leidenfrost states, *NASA Tech. Note. No D-3226.* [193]
Baumeister, K F, Graham, R W and Henry, R E (1973) Momentum flux in two-phase two-component, low quality flow. *ASME Symp. Heat Transfer Fundam. and Indust. Applic., Symp. Series 131,* **69**. [49,130]
Bayley, F J and Turner, A B (1968) Bibliography of heat transfer instrumentation. *Aeronautical Res. Council,* R and M No 3512. [30]
Bayoumi, M, Charlot, R and Ricque, R (1976) Determination of mass flow rate and quality distributions between the subchannels of a heated bundle. *European Two-Phase Flow Meeting, Erlangen, June, 1976,* Paper No D11, 14. [120]
Bearden, R g (1977) Progress on LOFT upgrade drag-disc turbine separate effect tests. US Nuclear Regulatory Commission, *Proceedings of Meeting of Review Group on Two-Phase Flow Instrumentation, January, 1977,* NUREG-0375 (Paper No I.5). [98]
Becker, H A, Hottel, H C and Williams, G C (1967) On the light-scatter technique for the study of turbulence and mixing. *Jnl. Fluid Mechs.* **30** (2), 259-283. [127]

Beckerleg, B A and Kiel, I (1970) An assessment of conductivity probes for the detection of two-phase flow regimes. *CEGB* Report No CEGB/RD/M/M68. [107]

Bedi, P S (1971) Simplified optical arrangement for the laser-doppler velocimeter, *Journal of Physics E: Scientific Instruments*, **4**, 27-28 [135]

Beer, H (1971) Interfacial velocities and bubble growth in nucleate boiling. *Int. Symp. Two-phase systems, Haifa, Israel, August, 1971*, Session 1, Paper 7. [142,192]

Begley, R F, Harvey, A B, Byer, R L and Hudson, B S (1974) A new spectroscopic technique: coherent anti-Stokes Raman spectroscopy. *American Laboratory*, November, 1974. [145]

Behar, M (1963) A delayed-flash stroboscope for bubble studies. *La Houille Blanche*. **18**, 692-696. [192]

Behar, M (1966) A delayed-flash stroboscope for bubble studies. *La Coin due Laboratoire*, Report No 6, 692-696. [192]

Behar, M and Semeria, R (1963a) Strioscopy applied to the study of boiling and degassing. *Soc. Hydrotech. France, 1st session, Comite Technique, March 1963*. [193,194]

Behar, M and Semeria, R (1963b) Strioscopic evidence of certain thermal transfer mechanisms in the degassing and boiling of water. *CR Acad. Sci., Paris*, **257**, 2801-2803. [194]

Belda, W (1975) Dryout delay through loss of coolant in nuclear reactors. *Doctor-Ingenieur Thesis, Technischen Universitat, Hannover*. [80]

Belda, W, Mayinger, F and Viert, K P (1974) Blow down studies with refrigerant R12. *Proceedings DER Reaktortagung des DAF, Berlin, 1974*. [203]

Bendick, P J (1974) A laser-doppler velocimeter to measure 'instantaneous' velocity profiles *In* "Flow - its measurement and control in science and industry", **I**, Part 2, 1033-1036, ISA. [137]

Benedict, R P (1969) "Fundamentals of temperature, pressure and flow measurements" Wiley. [22]

Benkert, J, Raes, K H, Reimche, W and Stegemann, D (1977) Studies of the dynamic behaviour of different detectors of gas and vapour bubbles in two-phase flows. *Kerntechnische Gesellschaft im Deutschen Atomforum ev*, Proceedings of a specialist conference, February-March 1977, Hannover; *In* "Experimental techniques in the field of thermodynamics and fluid dynamics. Part II; thermometry techniques and simulation of thermodynamics processes", 213 Interatom, Berg-Gladbach, West Germany. [76]

Benn, D and Shock, R A W (1974) An infra-red burnout detector. *AERE*-R7338. [67]

Bennett, A W, Hewitt, G F and Kearsey, H A (1965-66) Flow visualisation studies of boiling at high pressure. *Proc. Inst. Mech. Eng.* **180**, Part 3C, 1. [189]

Bennett, A W, Hewitt, G F, Kearsey, H A and Keys, R K F (1966) The wetting of hot surfaces by water in a steam environment at high pressure. *AERE*-R5146. [83,84]

Bennett, A W, Hewitt, G F, Kearsey, H A, Keys, R K F and Stinchcombe, R A (1969) Measurement of liquid film flow rates at 1000 psia in upward steam water flow in a vertical heated tube. *AERE*-R5809. [171,172]

Benson, J M (1974) Survey of thermal devices for measuring flow, In "Flow - its measurement and control in science and industry", I, Part 2, 549-554, ISA. [132]

Ben-Yosef, N, Ginio, O and Mahlab, D (1975) Bubble size distribution measurement by doppler velocimeter. *J. Appl. Phys*. **46** (2), 738-740. [160]

Berenson, P J and Stone, R A (1965) A photographic study of the mechanism of forced convection vaporization. *AIChE*, **61**, 213-219. [189]

Bergles, A E (1967) Critical heat-flux and flow-pattern observations for low-pressure water flowing in tubes. *J. Heat Transfer*. **89**, 69-74. [107]

Bergles, A E (1969) Electrical probes for study of two-phase flows. *Eleventh Nat. ASME/AIChE Heat Transfer Conf. Minneapolis*, 70-81. [22,113]

Bergles, A E, Clawson, L G, Roos, J P and Bourne, J G (1967) Investigation of boiling-flow regimes and critical heat-flux. *Dynatech Corpn*, Report No. NYO-3304-10 (Aug. 1966-March 1967). [107,114]

Bergles, A E and Roos, J P (1966) Film thickness and critical heat flux observations for high pressure water in spray annular flow. *Int. Symp. Res. Co-current Gas-Liquid Flow, Sept. 1968, Univ. Waterloo, Canada*, **2**, Paper B6.

Bergles, A E, Roos, J P and Bourne, J G (1968) Investigations of boiling flow regimes and critical heat flux. *NYO*-3304-13. [107,114]

Bergles, A E and Suo, M (1966) Investigation of boiling-water flow regimes at high pressure. *Dynatech Report No. NYO*-3304-8. [106,107]

Bertoletti, S, Lombardi, C and Silvestri, M (1964) Heat transfer to steam-water mixtures. *CISE, Milan*, Report No. R-78. [34]

Bestenbreur, T P and Spigt, C L (1967) Study of mixing between adjacent channels in an atmospheric air-water system. *Two-phase Flow Meeting, Winfrith, UK, June 1967*. [149]

Bexon, R (1973) Magnification in aerosol sizing by holography. *Journal of Physics E: Scientific Instruments*, **6**, 245-248. [152]

Bickel, L W and Keltner, N R (1973) Measuring transient surface temperatures, *Instruments and control systems*, 59-62. [35]

Birchenough, A and Mason, J S (1974) Laser anemometry measurements in a gas-solid suspension flow. *Opt. Laser Technol*. **8** (6), 253-259. [138,159]

Birchenough, A and Mason, J S (1976) Local particle velocity measurements with a laser anemometer in an upward flowing gas-solid suspension. *Powder Technol*. **14** (1), 139-152. [138,159]

Birchenough, A and Mason, J S (1977) Particle velocity and axial turbulence intensity measurements in a dilute gas-solid suspension flowing vertically upwards. *J. Powder Bulk Solids Technol*. **1** (1), 6-12. [185]

Bizon, F (1965) Two-phase flow measurement with sharp-edged orifices and venturis. *Atomic Energy Canada Ltd*. Report No AECL-2273. [96]

Blair, M F and Lander, R D (1974) New techniques for measuring film cooling effectiveness and heat transfer. *ASME* Preprint No 74-HT-8. [167]

Blake, K A, Jespersen, K I and Lynch, D (1971) A traversing laser velocimeter. *Optics and laser technol*. **3** (4), 208-210. [135]

Blatz, H (1959) "Radiation hygiene handbook". McGraw Hill, New York. [48]
Blows, L G and Tanner, L H (1974) A method for the measurement of fluid surface velocities, using particles and a laser-light source, *Journal of Physics E: Scientific Instruments*, **7**, 402-405. [131]
Board, S J (1969) Experimental observations of vapour bubble growth in various constrained geometries. *CEGB* Report No RD/B/M/1418. [77]
Bockh, P V and Chawla, J M (1972) The velocity of pressure-wave propagation in fluid-fluid systems. *European Two-Phase Flow Group Meeting, CNEN Rome, June, 1972.* Paper No C5. [71,72]
Bode, H (1972) Heat and mass transfer around growing vapour bubbles. *Warme-Und Stoffuber Tragung*, **5**, 134-140. [46]
Bolotov, A A, Vaisblat, M B and Minukhin, L A (1967) Research into the flow of mixtures of steam and liquids in vertical pipes. *Thermal Engng.* **14**, 113-119. [121]
Bonnet, C, Macke, F, Morin, R and Salomon, J (1966) Temperature fluctuations in the heating wall and in the boiling liquid during nucleate boiling. *Euratom Report, Ispra-Italy,* No EUR 3162F. [188]
Bonnet, J A (1971 and 1972) On the use of acoustical and optical methods to detect boiling in nucleate reactors. *U. Microfilms,* No 72-14811. [76]
Bonnet, J A and Osborn, R K (1971a) Neutronic-acoustic detection of the onset of bulk boiling in liquid sodium. *Nucl. Sci. Engng.* **45**, 314-320. [76]
Bonnet, J A and Osborn, R K (1971b) Incipient boiling detection in sodium-filled fast reactors. *Amer. Nucl. Soc. Trans.* 309-310. [76]
Boothroyd, R G (1967a) An anemometric isokinetic sampling probe for aerosols. *J. Sci. Instr.* **44**, 249-253. [118,149]
Boothroyd, R G (1971) "Flowing gas-solid suspensions". Book, Chapman and Hall Ltd, London. [22]
Borgartz, B O, O'Brien, T P, Rees, N J M and Smith, A V (1969) Experimental studies of water depressurisation through simple pipe systems. *Meeting on Depressurisation Effects in Water Cooled Power Reactors, Battelle Inst. Frankfurt, 1969.* [72,73,75]
Bosio, J, Malnes, D and Rolstade, E (1970). Noise in two-phase flow. A short survey of the topic including papers presented at the European two-phase flow meeting in Karlsruhe, June 1969. *European Two-Phase Flow Group Meeting, Milan, June, 1970.* [76]
Bossel, H H, Hiller, W J and Meier, G E A (1972) Noise-cancelling signal difference method for optical velocity measurements, *Journal of Physics E. Scientific Instruments*, **5**, 893-896. [137]
Bossel, H H, Hiller, W J and Meier, G E A (1972) Self-aligning comparison beam methods for one-two-and three-dimensional optical velocity measurements, *Journal of Physics E: Scientific Instruments*, **5**, 897-900. [137]
Botterill, J M S and Sealey, C J (1970) Radiative heat transfer between a gas-fluidised bed and an exchange surface. *Brit. Chem. Engng.* **15** (9), 1167-1168. [38]
Bouche, R R (1967) Understanding accelerometers. *Electronic Engr.* **26** (4), 90-94. [86]
Bolman, H, van Koppen, C W J and Raas, L J (1974) Some investigations on the influence of the heat flux on flow patterns in vertical

boiler tubes. *European Two-Phase Flow Group Meeting, Harwell, June 1974*, Paper A2. [58]

Boure, J A (1972) Modelling methods in two-phase flow thermohydraulics, *Meeting on Two-Phase Thermohydraulics, CNEN Rome, June 1972*. Invited lecture No 7. [204]

Boure, J A (1975) Accident analysis, flow instability transient two-phase flow. A review. *European Two-Phase Flow Group Meeting, Haifa, Israel*. [69]

Boure, J A, Bergles, A E and Tong, L S (1973) Review of two-phase flow instability. *Nucl. Eng. Design*. **25**, 165. [91]

Boyd, L R (1959) Ion chamber can detect nucleate boiling. *Nucleonics*, **17** (3), 96-102. [77]

Bradfield, W S (1966) Liquid-solid contact in stable film boiling. *Ind. and Eng. Chem. Fundamentals*. **5**, Report No 2, 200-204. [192]

Brambilla, A, Rossi, G and Zanelli, S (1974) Characteristics of gas-liquid interface in vertical cocurrent two-phase flow by correlation analysis. *Quaderni Dell'ing. Chimico Italiano, Suppl. De La Chimica E Industria*, **4**, 49-54. [177]

Breugel, J W, van Stein, J J M and de Vries, R J (1969-70) Isokinetic sampling in a dense gas-solids stream on two-phase flow systems. *Proc. Instn. Mech. Engrs*. **184**, Pt 3C, 18-23. [118]

Briller, R and Robinson, M (1969) A method for measuring particle diffusivity in two-phase flow in the core of a duct. *AIChEJ*, **15**, 733-735. [149]

Brixy, H G (1971) Temperature measurement based on thermal noise. *Microfiche* No LA-TR-71-71. [167]

Brockett, G F and Johnson, R T (1976) Single phase and two-phase flow measurement techniques for reactor safety studies. *EPRI*-NP-195. [22,27,28,34,69]

Brook, A J, Kaiser, A and Peppler, W (1977) Flow rundown experiments in a seven pin bundle. *Nucl. Eng. Des*. **43** No 2, 273-283. [77]

Brown, D J, Jensen, A and Whalley, P B (1975) Non-equilibrium effects in heated and unheated annular two-phase flow. *ASME* 75 WA/HT 7. [171,177]

Brown, M (1974a) The measurement of two-phase flow parameters in an evaporator tube in an operating boiler. *Inst. Chem. E. Symp. Ser*. **1** (38). [99]

Brown, M (1974b) Measurement of mass flow-rates of water and steam in evaporator tubes on operating boilers using radiotracers. *Int. J. Appl. Radiation Isotopes*, **25** (7), 289-294. [99]

Bruce, J M (1972) Fluid-to-fluid modelling criteria for two-phase pressure drop. *AECL Report* AECL-4263. [206]

Bruun, H (1971) Linearisation and hot-wire anemometry, *Journal of Physics E: Scientific Instruments*, **4**, 815-820. [128]

Buchberg, H, Romie, F, Lipkis, R and Greenfield, M (1951) Final report on studies in boiling heat transfer. *Univ. California*, Contract No AT-11-1-GEN-9. [51]

Burick, R J (1972) Probe measures gas and liquid mass-flux in high mass-flow ratio two-phase flows. *NASA Tech. Brief* No B72-10546. [118]

Burick, R J, Scheuerman, C H and Falk, A Y (1971) Determination of local values of gas and liquid mass flux in highly loaded two-phase flow, *Symposium on Flow - its Measurement and Control in*

Science and Industry, Pittsburgh, May, 1971. [117,128]
Burick, R J, Scheuerman, C H and Falk, A Y (1974) Determination of local values of gas and liquid mass flow in highly loaded two-phase flow. *In* "Flow: its measurement and control in Sci. & Indust". *Instr. Soc. America,* **1** (1), 153-160. [120]
Burnett, P and Burns, J S (1977) The influence of lead temperature on the accuracy of various stainless-steel sheath, mineral-insulated nickel-chromium/nickel aluminium thermocouplks. *UKAEA* report ND-R-36(W). [32,33]
Burton, E J (1974) Acoustic diagnostics and nuclear power plants. *J. Brit. Nucl. Ener. Soc.* **13** (2), 183. [76]
Butterworth, D (1968) Air-water climbing film flow in an eccentric annulus. *AERE*-R5787. Also: *Internat. Symp. Res. Cocurrent Gas-Liquid Flow. Sept. 1968, Univ. Waterloo.* Paper B2. [112,114,172]
Butterworth, D, Hazell, F C and Pulling, D J A (1974) A technique for measuring local heat transfer coefficients in a horizontal tube. *Symp. Multiphase Flow Systems, Univ. Strathclyde, 1974. I. Chem. E. Symp. Series* No 38. [172]
Butterworth, D and Hewitt, G F (Eds) (1977) "Two phase flow and heat transfer", Oxford University Press, Oxford. [1,7,11,12,13]
Butterworth, D and Owen, R G (1975) The quenching of hot surfaces by top and bottom flooding: A review. *AERE*-R7992. [83]
Butterworth, D and Pulling, D J (1973) Film flow and film thickness measurements for horizontal annular air-water flow. *AERE*-R-7576. [37,172,173]
Butterworth, D and Pulling, D J (1974) Mechanisms in horizontal annular air-water flow. *M.E. and C.E. Symp.* **1**, A2.

Cadek, F F, Dominicis, D P and Leyse, R H (1971) PWR FLECHT final report. *WCAT*-7665. [166]
Calderbank, P H (1958) Physical rate processes in industrial fermentation. Part 1. The interfacial area in gas-liquid contacting with mechanical agitation. *Trans. Instn. Chem. Engrs.* **36**, 443-463. [154]
Calderbank, P H and Johnson, D S L (1970) Mechanics and mass transfer of single bubbles in free rise through newtonian and non-newtonian liquids. *Chem. Eng. Sci.* **25**, 235-255. [42]
Campanile, A, Galimi, G and Goffi, M (1969) Method of investigating the motion of gas bubbles in liquids by means of a radioactive bubble. *Energ. Nucleare,* **16** (6), 386-391. [133]
Campanile, A and Pozzi, G P (1972) Low rate emergency reflooding heat transfer tests in rod bundle. *Crest Specialist meeting on emergency core cooling for light water reactors, Munich,* MRR 115, 1. [82]
Carey, W M, Albrecht, R W (eds) (1976). Potential for LMFBR boiling detection by acoustic/neutronic monitoring. *Proc. Seminar Argonne Nat. Lab. April 1976. (Microfiche* No ANL-CT-76-34). [76]
Carlos, C R and Richardson, J F (1967) Particle speed distribution in a fluidised system. *Chem. Engng. Sci.* **22**, 705-706. [189]
Carrard, G and Ledwidge, T J (1971) Measurement of slip distribution and average void-fraction in an air-water mixture. *Int. Symp. on Two-Phase Systems, Haifa, Israel, Aug-Sept 1971.* Paper 3-13. [57]
Carzaniga, R, Holtbecker, H, Premoli, A and Soma, E (1969) Preliminary experiments on the blow-down of a pressure vessel. *Crest Specialist Meeting, depressurisation effects in water-cooled reactors,*

Frankfurt, 1969. [72]

Castellana, F S and Casterline, J E (1972) Subchannel flow and enthalpy distributions at the exit of a typical nuclear fuel core geometry. *Nucl. Engn. Design.* **22** (1), 3-18. [120,149]

Castellani, F, Curzio, G and Pieve, L (1975) Experimental tests on the applicability of a radioactive tracer technique for measuring coolant mixing in nuclear reactor fuel subassemblies. *Nucl. Engng. Design.* **32** (1), 121-129. [149]

Catling, E, Cartwright, D K and Willis, P (1977) A sodium-boiling noise generator for acoustic commissioning of PFR. *UKAEA* Report No-R-61(R). [76]

Ceelen, D, Gebureck, P and Stegemann, D (1976) Determination of steam-bubble velocity and steam void fraction in BWR fuel elements by stochastic methods. *Atomkernenergie,* **27** (4), 239-242. [64,141]

Cedolin, L, Hassid, A and Rossini, T (1970) Vibrations measurement in a 19-rod cluster fuel element with two-phase (gas and liquid) coolant flow. *European Two-Phase Flow Group Meeting, Milan, June 1970,* Paper C4. [86]

Cedolin, L, Hassid, A and Rossini, T (1971) Vibrations induced by two-phase (gas and liquid) coolant flow in the power channels of a pressure-tube-type nuclear reactor. *European Two-Phase Flow Group Meeting, RISO, June 1971.* Paper A2. [86]

CGS Thermodynamics (1974) "The thermocouple handbook". [32]

Charvonia, D A (1962) An experimental investigation of the mean liquid film thickness and the characteristics of the interfacial surface in annular, two-phase flow. *ASME* Paper No 61-WA-243. [178]

Chaudhry, A B, Emerton, A C and Jackson, R (1965) Flow regions in the co-current upwards flow of water and air. *Symp. Two-Phase Flow, Exeter, 1965.* Paper B2. [27]

Chedville, C, Lions, N and Rosse, M (1969) Method and device for gas content of a flowing two-phase mixture. *Patent Specification No* 1,138,433. [62]

Chen, J C, Delale, C, Eberhardt, N et al (1977) Two-phase flow instrumentation research. *Paper presented NRC 5th Water Reactor Safety Res. Informatiln Meeting, Gaithersburg, Maryland, USA. Nov. 1977.* [122,166,167]

Chen, J, Kalish, S and Schoener, G A (1968) Probe for detection of voids in liquid metals. *Rev. Sci. Instrum.* **39**, 1710-1713. [122]

Chesters, A K (1975) The applicability of dynamic-similarity criteria to isothermal, liquid-gas, two-phase flows without mass transfer. *Int. J. Multiphase Flow,* **2** (2), 191-212. [205]

Chesters, A K (1977) A note on the centrifugal scaling of horizontal, isothermal, liquid-gas flows without mass transfer. *Int. J. Multiphase Flow,* **3** (3), 235-241. [205]

Chisholm, D (1967) A theoretical basis for the Lockhart-Martinelli correlation for two phase flow. *National Engineering Laboratory, UK.* Report No 310. Also published in *Int. J. Heat Mass Transfer,* **10**, 1767. [11]

Chisholm, D (1972) The compressible flow of two-phase mixtures through orifices, nozzles and venturimeters. *Nat. Engng. Lab. UK,* Report No 549, 66-79. [96]

Chisholm, D and Sutherland, L A (1969-70) Prediction of pressure gradients in pipeline systems during two phase flow. *Proc. Inst.*

Mech. Eng. **184** (3C), 24. [11]
Chojnowski, B and Wilson, F W (1972) Critical heat flux for large bore boiler tubes. *CEGB* Report No CEGB/R/M/N652. [66]
Chu, K J and Dukler, A E (1974) Statistical characteristics of thin, wavy films: Part II. Studies of the substrate and its wave structure. *AIChE*, **20** (4), 695-706. [112,175]
Chu, K J and Dukler, A E (1975) Statistical characteristics of thin, wavy films, *AIChEJ* **21** (3), 583-593. [175,185]
Chung, H S and Murgatroyd, W (1967) Interfacial disturbances and stable waves in two-phase annular flow. *Symp. Two-Phase Flow Dynamics, Eindhoven, Sept. 1967*. Session X, 10.4. [176]
Cimorelli, L, di Bartolomeo, M and Premoli, A (1968) Measurement of volume fraction of vapour in a boiling channel by the impedance method. *CNEN*, Report RT/ING (65)7, (1965). Translated by Crozy, A, UKAEA, *AERE Trans*. No LB/G/2697. [57]
Cimorelli, L and Evangelisti, R (1967) The amplification of the capacitance method for void-fraction measurement in bulk-boiling conditions, *Int. Jnl. Heat Mass Transfer*. **10**, 277-288. [57]
Cimorelli, L and Evangelisti, R (1969) Experimental determination of the slip ratio in a vertical boiling channel, under adiabatic conditions of atmospheric pressure. *Int. J. Heat Mass Transfer*. **12**, 713-726. [57]
Clark, C L (1963) "Encyclopaedia of X-rays and gamma rays. Rhienhold. [51]
Clark, D L, Heddleston, F A and Lawson, C G (1974) Requirements for heaters for blow-down heat transfer tests in pressurised-water reactor environments. *ORNL*-TM-4644. [67]
Clark, N O (1946) A meter for the measurement of the properties and quantity of foam. *Jnl. Scient. Instrum.* **23**, 256-259. [98]
Class, G, Fiege, A and Fischer, M (1972) Investigations on fuel rod failure during a loss-of-coolant accident in consideration of ECCS - criteria. *Crest Specialist Meeting on emergency core cooling for light water reactors, Munich*, MRR 115, **1**. [78]
Claverie, J (1970) Measurements in wet steam. *Conference on Two-Phase Flow in a Vapour Generator. Ermenonville, Sept. 1970*. [96]
Clayton, C G (1960), Precise tracer measurement of liquid and gas flow. *Nucleonics*, **18** (7), 96-100. [99]
Clayton, C G (1964) The measurement of flow of liquids and gases using radioactive isotopes. *J. Brit. Nucl. Energy. Soc.* 252-268. [33]
Collier, J G (1972) "Forced convective boiling and condensation" McGraw-Hill. [1,11]
Collier, J G (1975) Two-phase flow and heat transfer in water-cooled nuclear reactors. *Course at Dartmouth College, Round Table discussion on measurement techniques, Aug. 1975*. [69]
Collier, J G and Hewitt, G F (1960) Data on the vertical flow of air-water mixtures in the annular and dispersed flow regions. Part B. Film thickness and entrainment data and analysis of pressure-drop measurements. *AERE*-R3455. [169]
Collier, J G and Hewitt, G F (1964) Film thickness measurements. *AERE*-R4684. [112]
Collier, J G and Hewitt, G F (1966) Experimental techniques in two-

phase flow. *Brit. Chem. Engng.* **11** (12), 1526-1531. [189]
Collingham, R E and Firey, J C (1963) Velocity of sound measurements in wet steam. *I and E.C. Proc. Des. and Dev.* **2** (3), 197-202. [71]
Collins, D B and Gacesa, M (1970) Measurement of steam quality in two-phase upflow with venturimeters and orifice plates. *Microfiche* AED-Conf. 072-002. [96]
Colombo, A, Hassid, A and Premoli, A (1967) Density measurements in heated channels at high pressure by means of a quick-closing valve method. *CISE Milan*, Report No CISE-R-232. Also *Energ. Nucleare*, **15**, 119-128. [61]
Colombo, A and Premoli, A (1972) An experimental investigation on the use of elbows as two-phase quality meters. *European Two-Phase Flow Group Meeting, CNEN Rome, June 1972.* Paper No D9. [97]
Comorelli, L and Premoli, A (1966) Measurement of void fraction with impedance gauge technique, *Energia Nucleare*, **13** (1), 12-23.
Condon, R A and Sher, N C (1961) Measurement of void fraction in parallel rod arrays. *Allis Chalmers Manuf. Comp. Report.* [49]
Coney, M W E (1971) The variation of liquid film thickness in horizontal two-phase flow. *CEGB* Report No CERL/RD/L/N/ 273/71. [112]
Coney, M W E (1972) The application of conductance probes to the measurement of liquid film thickness in two-phase flow. *CEGB* Rep. CERL/RD/L/N 231/72. [112,113]
Coney, W E and Fisher, S A (1976) Instrumentation for twp-phase flow in use or under development at the Central Electricity Research Laboratories. *Paper presented at the meeting of the European Two-Phase Flow Group, Erlangen, Germany, June 1976.* [172,173]
Connors, H J (1970) Fluid elastic vibration of tube arrays excited by cross-flow, *Paper presented at Winter Annual Meeting, ASME, New York, December 1970.* [85]
Cook, R A and Clark, R H (1971) The experimental determination of velocity profiles in smooth falling liquid films. *Can. J. Chem. Engng.* **49** (3), 412-416. [131,184]
Cooper, K D, Hewitt, G F and Pinchin, B (1963) Photography of two-phase flow, *AERE-R4301*. [105,152,189]
Cooper, M G, Judd, A M and Pike, R A (1978) Shape and departure of single bubbles growing at a wall. *Proceedings of the 6th International Heat Transfer Conference, Toronto, August, 1978.* **1**, 115. [192]
Cooper, M G and Lloyd, A J P (1966) Transient local heat flux in nucleate boiling. *Proc. 3rd Int. Heat Trans. Conf. Chicago, Aug. 1966.* **3**, 193-203. [188]
Corradi, F (1977) New instruments and methods for high precision thermocouple and platinum resistance thermometry. *Kerntechnische Gesellschaft im Deutschen Atomforum ev*, Proceedings of a specialist conference, Feb-March 1977, Hannover; "Experimental techniques in the field of thermodynamics and fluid dynamics. Part II; thermometry techniques and simulation of thermodynamics processes" 345, Interatom, Berg-Gladbach, West Germany. [32,33]
Costello, C P (1959) Aspects of local boiling effects on density and pressure drop. *ASME Paper* No 59-HT-18. [54]
Cousins, L B, Denton, W H and Hewitt, G F (1965) Liquid mass transfer in annular two-phase flow. *AERE-R4926*, 1965: Also *Symp. Two-Phase Flow, Exeter, June 1965.* Paper C4. [146,150,170]

Cousins, L B and Hewitt, G F (1968a) Liquid-phase mass transfer in annular two-phase flow: droplet deposition and liquid entrainment. *AERE-R*5657. [127,150]

Cousins, L B and Hewitt, G F (1968b) Liquid-phase mass-transfer in annular two-phase flow: Radial liquid mixing. *AERE*-R5693. [146,147]

Crane, R I and Melling, A (1975) Velocity measurements in wet steam flows by laser anemometry and pitot tube. *J. Fluids Eng.* **97**, Series 1, No 1, 113-116. [138]

Crane, R I and Moore, K J (1972) Interpretation of pitot pressure in compressible two-phase flow. *J. Mech. Eng. Sci.* **14** (2), 128-133 [129]

Cravarolo, L et al (1961) A beta-ray attenuation method for density measurements of liquid-gas mixtures in adiabatic flow. *Energia Nucleare*, **8**, 751-757. [54]

Cravarolo, L, Georgina, A, Hassid, A and Pedrocchi, E (1964) A device for the measurement of shear stress on the wall of a conduit. Its application in the mean density determination in two-phase flow. Shear stress data in two-phase adiabatic vertical flow. *CISE Milan*, Report No R-82. [161,162,165]

Cumo, M, Ferrari, G and Farello, G E (1971) A photographic study of two-phase highly dispersed flows. *CNEN Italy*. Report No RT/ING (71), **8**. [189]

Currie, R J and Rhodes, E (1971) Non-disturbing flow visualisation in the liquid at a mobile sheared gas-liquid interface. *Int. Symp. Two-Phase Systems, Haifa, Israel, Aug-Sept. 1971.* Session 2. [184]

Cybula, G J (1971) An investigation of the measurement of void fraction in air-water mixtures by the electrical impedance method. *Australian AEC* Report No AAEC/TM592. [57]

Cumo, M, Macica, C, Moronesi, M et al (1974) Modelling technqiues for the reproduction of the effect of burnout of water using freon (in Italian) *CNEN* Report No RT/ING. **6**. [204]

Dalmon, J and Lowe, H J (1957) Effect of dust-burden on pitot-static readings. *2nd pulverized fuel Conf*. Paper 6. [128]

Dandlouker, R and Iten, P D (1974) Direction sensitive laser-doppler velocimeter with polarised beams. *Applied Optics*. **13** (2), 286-290. [137]

Danel, F (1971) Cavitation research: The measurement of gas content in liquids. *La Houille Blanche*, **4**, 309. [127]

Danel, F and Delhaye, J M (1971) An optical probe for measuring local voidage in two-phase flow. *Mesures, Aug-Sept 1971.* 99-101. [125]

Davies, G A, Mojtehedi, W and Ponter, A B (1971) Measurement of contact angles under condensation conditions. The prediction of dropwise-filmwise transition. *Int. Jnl Heat Mass Transfer*, **14** 709-713. [174]

Davies, W E R and Unger, J I (1973) Velocity measurements in bubbly two-phase flows using laser-doppler anemometry (Pt 2) *Microfiche* No. N73-18506. [138]

Davis, E J (1969) Interfacial shear measurement for two-phase gas-liquid flow by means of Preston tubes. *Ind. Eng. Chem. Fund.* **8**, 153-159. [165]

Davis, E J and Cooper, T J (1969) Thermal entrance effects in

stratified gas-liquid flow: Experimental investigation. *Chem. Eng. Sci.* **24**, 509-520. [114,131]
Davis, G E (1955) Scattering of light by an air bubble in water. *Jnl. Opt. Soc. Amer.* **45** (7). [63,154]
Dawson, D A and Trass, O (1972) Mass transfer at rough surfaces. *Int. J. Heat Transfer,* **15**, 1317-1336. [46]
Deane, C W and Rohsenow, W M (1969) Mechanism and behaviour of nucleate boiling heat transfer to the alkali liquid metals. *Mass. Inst. Technol.* Report No DSR 76303-65. [188]
Deckwer, W D, Burckhart, R and Zoll, G (1974) Mixing and mass transfer in tall bubble columns. *Chem. Eng. Sci.* **29** (11), 2177-2188. [41]
Deich, M E, Abromov Yu, I, Shanin, V K and Khizanashvili, M D (1974) Experimental investigation of the slip of the liquid phase at the outlet of nozzle blade cascades in turbines. *Therm. Engng.* **21** (6), 66-72. [139]
Deich, M E, Saltanov, G A and Kurshakov, A V (1971) Investigating the kinetics of phase transitions in shock waves in wet steam flow. *Thermal Engng.* **18** (4), 127-131. [154]
Deich, M E, Tsiklauri, G W and Shanin, V K (1972) Investigation of flows of wet steam in nozzles. *High Temp.* **10** (1), 102-107. [154]
Delhaye, J M (1966) Behaviour of a pitot tube in two-phase flow. *CEN Grenoble,* Report TT No 220. [128]
Delhaye, J M (1968a) Measurement of the local void fractions in two-phase air-water flow with a hot film anemometer. *CEA*-R-3465(E). [125]
Delhaye, J M (1968b) Anemometry with hot film in two-phase flow. Measure of levels in presence of gas. *CR Acad. Sci. Paris.* **266**, Series A, 370-373. [125]
Delhaye, J M (1969) Hot-film anemometry in two-phase flow. *Eleventh Nat. ASME/AIChE Ht. Transfer Conf. Minneapolis, Aug. 1969,* 58-69. [125]
Delhaye, J M (1970) Instrumentation for two-phase flow. *Conf. Two-Phase Flow in a Vapour Generator, Ermenonville, Sept. 1970.* [187]
Delhaye, J M (1972) Local measurements in two-phase flow. Fluid-dynamic measurements in the industrial environments. *Conference papers,* **1**, 191-200. [22,183]
Delhaye, J M (1974) Two-phase flow measurements. *Bull. D'Inform. Sci. Tech.* **197**, 5-20. [22]
Delhaye, J M and Achard, J L (1977) On the use of averaging operators in two-phase flow modelling. *Symposium on the thermal and hydraulic aspects of nuclear reactor safety. Volume 1: light water reactors,* (Ed O C Jones and S G Bankoff), American Society of Mechanical Engineers. [22,175]
Delhaye, J M and Chevrier, C (1966) The use of resistivity probes for the measurement of local void fraction in two-phase flow. *CEN Grenoble,* Report No TT 70. [122,123]
Delhaye, J M, Galaup, J P, Reocreux, M and Ricque, R (1973a) Metrology of two-phase flow - different methods. *CEN Grenoble,* Report No. CEA-R-4457. [22]
Delhaye, J M and Galaup, J P (1974) Measurement of local void fraction in Freon 12 with a 0.1 mm optical fibre probe. *European two phase flow meeting, June 1974, Harwell. Paper* A6. [125]
Delhaye, J M and Jones, O C (1975) Measurement techniques for

transient and statistical studies of two-phase, gas liquid flows. *Internat. J. Multiphase Flow.* [175,186]

Delhaye, J M, Semeria, R and Flamand, J C (1972) Measurement with a micro-thermocouple of the local void-fraction and the temperature of the liquid and vapour in two-phase flow with phase change. *CEA-R-4302.* [186]

Delhaye, J M, Semeria, R and Flamand, J C (1973b) Void fraction, vapour and liquid temperatures: local measurements in two-phase flow using a microthermocouple. *J. Heat Transfer*, 365-370. [186, 187]

Dement'ev, B A and Skachek, M A (1973) Determination of the true steam contents in cross sections of a pipe of large diameter. *Thermal Engng.* **20**, No 9, 75-90. [53,121]

Denham, M K, Briard, P and Patrick, M A (1975) A directionally-sensitive laser anemometer for velocity measurements in highly turbulent flows. *Journal of Physics E: Scientific Instruments.* **8**, 681-685. [137]

Dennis, J A (1957) Investigation of the possibility of using a neutron source to measure the specific gravities of steam-water mixtures in a thick pipe, *AERE*-EL/M-97. [54]

Dent, J C and Derham, J A (1972) A method for checking the consistency of velocity computations from hot wire anemometer measurements in variable density flows. *Journal of Physics E: Scientific Instruments,* **5**, 271-273. [128]

Deprisco, C F, Kartluke, H, Maropis, N and Tarpley, W B (1962) Ultrasonic detection of incipient boiling and cavitation. *Aeroprojects Inc.* NY 10010-UC 37. [77]

Derauz, R (1967) Determination of the rate of two-phase air-water flow by measurement of electrical conductivity. *CEN Grenoble,* Note TT 281. [57]

Derbyshire, R T P, Hewitt, G F and Nicholls, B (1969) X-radiography of two-phase gas-liquid flow. *AERE*-M1321. [106]

Dickson, A N and Wood, J D (1972) Two-phase flow through a sharp-edged orifice. *Nat. Engng. Lab, UK.* Report No 549, 36-47. [96]

Dijkman, F (1970) Stability aspects of a boiling channel with a sine-shaped heat flux. *4h Internat. Heat Transfer Conf. Paris - Versailles, 1970,* **5**, B4.3. [92]

Dix, G E (1971) Vapour void fractions for forced convection with subcooled boiling at low flow rates. *G.E. California,* Report No NEDO-10491. [125]

Dixon, A D (1957) Gamma absorption techniques applied to steam dryness determinations, *IGR*-TN/W-489. [47]

Doe, P E (1967) A new method of measuring humidity in a small space. *Int. J. Heat Mass Transfer*, **10**, 311-319. [167]

Downing, P M (1972) Reverse flow sensing hot wire anemometer. *Journal of Physics E: Scientific Instruments,* **5**, 849-851. [128]

Dsarasov Yu, I, Kolchugin, B A and Liverant, E I (1974) Experimental study of relation between heat and mass transfer characteristics in evaporating channels. *5th Int. Heat Transfer Conf. Tokyo, Sept. 1974,* **4**, Paper No B5.3, 195-199. [114,118]

Du Bousquet, J L (1968) Experimental study of mixing and cross-flow with a variable gap two subchannels. Test Section. *European Two-Phase Flow Group Meeting, Oslo, June 1968.* Paper A5. [149]

Duffey, R B and Porthouse, D T C (1972) Experiments on the cooling of high-temperature surfaces by water jets and drops. *Crest Specialist Meeting on Emergency Core Cooling for Light Water Reactors, Munich 1972.* MRR 115, **1**. [83]

Duffey, R B and Porthouse, D T C (1973) The physics of rewetting in water reactor emergency core-cooling. *Nuclear Eng. Des.* **25**, 379. [83,84]

Duke, E E and Schrock, V E (1961) Void volume, site density and bubble size for subcooled nucleate pool boiling. *Proc. of the 1961 Ht. Transfer and Fluid Mech. Inst.* 130-137. [61]

Dukler, A E and Bergelin, O P (1952) Characteristics of flow in falling liquid films. *Chem. Engng. Progr.* **14**, 557-563.

Dukler, A E and Hubbard, M G (1975) A model for gas-liquid slug flow in horizontal and near horizontal tubes. *Private communication.* [185]

Durst, F, Melling, A and Whitelaw, J H (1976) "Principles and practice of laser-Doppler anemometry" Academic Press, London and New York. [135]

Durst, F and Whitelaw, J H (1971) Integrated optical units for laser anemometry, *Journal of Physics E: Scientific Instruments.* **4**, 804-808. [135]

Durst, F and Zare, M (1975) Laser-Doppler measurements in two-phase flow. *Univ. Karlsruhe*, SFB-80/TM/63. [138,185]

Durst, F and Zare, M (1976) Laser doppler measurements in two-phase flows. *Symp. Two Phase Flow Cavitation in Power Generation Systems, Grenoble, March-April 1976.* 409-418. [138]

Dussourd, J L and Shapiro, A H (1955) A deceleration probe for measuring stagnation pressure and velocity of a particle-laden gas stream. *Mass. Inst. Technol.* Report, Contract N5OM-07878. [128]

D'Yachevskiy, A P, Leonchik, B I and Mayakin, V P (1970) Temperature measurement in a particle-conveying gas flow. *Heat Transfer Soviet Research*, **2** (6), 110-113. [188]

Dzakowic, E B and Dix, R C (1969) A rapid response impact tube in two-phase flow. *AIChE Jnl.* **15**, 466. [128]

Early, B (1976) "Practical instrumentation handbook" Scientific ERA Publications. [32]

Eckert, E R G and Goldstein, R J (1970) Measurement techniques in heat transfer. Published by Advisory Group for Aerospace Research and Development of NATO, AGARDograph No 130. [22]

Eddington, R B (1967) Investigation of supersonic shock phenomena in a two-phase (liquid-gas) tunnel. *JPL* Tech. Report 32-1096. [71]

Edwards, F J, Alker, C J R and Harvey, B (1974) A non-contact method of measuring surface temperature. *Fifth Internat. Ht. Transfer Conf. 1974.* Paper MA 2.4. [37]

Edwards, A R, Mather, D J, Flanagan, S et al (1977) Some limitations on the use of the assumption of thermal equilibrium between phases: A comparison between theory and experiments on simple water filled systems. *Conf. on Heat and Fluid Flow in Water Reactor Safety, Manchester, Sept. 1977.* 27-32. [72]

Einav, S and Lee, S L (1973) Measurement of velocity distributions in two-phase suspension flows by the laser-Doppler technique. *Rev. Sci. Instrum.* **44** (10), 1478-1480. [138]

References

Elias, E and Yadigaroglu, G (1976) The reflooding phase of the LOCA - state of the art II. Rewetting and liquid entrainment. *Nato Advanced Study Inst. Two Phase Flows Heat Transfer, Istanbul, Turkey, Aug. 1976.* [83]

Elliot, D F and Rose, P W (1971) The quenching of a heated zircaloy surface by a film of water in a steam environment at pressures up to 53 bar. *AEEW*-M1027. [83]

Elliott, L C and Dukler, A E (1970) Condensation in a concentric annulus in the presence of a non-condensing gas. Univ. Houston. *Private communication.* [35,36]

Ellis, A T (1966) On jets and shock waves from cavitation. *Symp. on Naval Hydrodynamics, Washington 1976.* ACR-146, 137-154. [77]

English, D, Blacker, P T and Simmons, W E (1955) Boiling and density studies at atmospheric pressure. *AERE* BO/M20. [54]

Engler, C and von Holzen, G (1977) Fitting of thermocouples into the cladding walls of electrically heated experimental fuel pins. *Kerntechnische Gesselschaft im Deutschen Atomforum ev,* Proceedings of a Specialist Conference, Feb-March 1977, Hannover; "Experimental techniques in the field of thermodynamics and fluid dynamics. Part II; thermometry techniques and simulation of thermodynamics processes". 115, Interatom, Berg-Gladbach, West Germany. [80]

Epstein, H C (1972) Development and application of a 0.14gm piezoelectric accelerometer. *Proc. 27th Ann. Conf. and Exhibit Instrument Soc. Amer.* Paper No 72-619.

Era, A , Gaspari, G P, Hassid, A, Milani, A and Zavattarelli, R (1966) Heat transfer data in the liquid-deficient region for steam/water mixture at 70 kg/cm^2, flowing in tubular and annular circuits. *CISE (Milan)* Report R-184. [83]

Eriksson, S O (1975) DSF-P1: An emergency core-cooling experiment for BWR's trans. *ANS.* **20**, 494. [83]

Etherington, H (1958) "Nuclear engineering handbook" McGraw Hill, New York. [48]

Evangelisti, R and Lupoli, P (1969) The void fraction in an annular channel at atmospheric pressure. *Int. Jnl. Heat Mass Transfer,* **12**, 699-711 also *Ingegn. Nucleare.* **10** (2), 19-36. [49]

Evans, G V (1970) Evaluation of a radiotracer method of measuring mean liquid velocities in an air-water flow system. *AERE*-R11/103/RS. [133]

Evans, G V, Robertson, J M and Spackman, R (1971) Evaluation of a radiotracer method of measuring mean liquid phase velocities in air-water, two-phase flow. *AERE*-R6753. [62]

Evans, R D (1963) Gamma rays. *In* "American Institute of Physics Handbook" McGraw Hill. [51]

Evans, R G, Gouse, S W and Bergles, A E (1970) Pressure wave propagation in adiabatic slug-annular-mist two-phase, gas-liquid flow. *Chem. Engng. Sci.* **25**, 569-581 [71]

Farmer, R, Griffith, P and Rohsenow, W M (1970) Liquid droplet deposition in two-phase flow. *Jnl. Heat Transfer.* **92**, Series C, 587-594. [150]

Farmer, W M (1972) Measurement of particle size, number density, and velocity using a laser interferometer. *Appl. Opt.* **11** (11), 2603-2612. [138,150,157,158]

Farmer, W M and Hornkohl, J O (1973) Two-component, self-aligning laser vector velocimeter, *Applied Optics.* **12** (11), 2636-2640. [137]

Farukhi, M N and Parker, J D (1974) A visual study of air-water mixtures flowing inside serpentine tubes. *5th Internat. Ht. Transfer Conf. Sci. Council, Japan.* **4**. [189]

Fauske, H K (1965) Two-phase, two-and one-component critical flow. *Proc. Symp. Two-Phase Flow, Exeter, 1965.* Paper G1. [49]

Fauske, H K (1976) Two-phase flow and heat transfer problems of interest in liquid metal fast breeder reactor safety analysis. *NATO Advanced Study Inst. Two Phase Flows Heat Transfer, Istanbul, Turkey, Aug. 1976.* [78]

Feinauer, E and Brockett, G F (1963) Some requirements for pressure instrumentation in evaluation of reactor incidents, reactor kinetics and control. TID-7662, *Proceedings of USAEC Symp., Univ. Arizona,* 491-500 (March 25-27 1963). [27]

Feldman, E M and Naff, S A (1975) Error analysis for 1-1/2 loop semi-scale system isothermal test data. *Aerojet Nuclear Company.* ANCR-1188. [22]

Feldman, C L, Nydick, S E and Kokernak, R P (1971) The speed of sound in single component, two-phase fluids - theoretical and experimental. *Int. Symp. Two-Phase Systems, Haifa.* [71]

Fenton, D L and Stukel, J J (1976) Measurement of local particle concentration in fully turbulent duct flow. *Int. J. Multiphase Flow.* **3**, 141. [127]

Ferrell, J K and Bylund, D M (1966) Low pressure steam water flow in a heated vertical channel. *USAEC. U.N. Carolina, Raleigh.* Final report No TID-23394. [49,107]

Ferrell, J K and McGee, J W (1965) An accurate one-shot gamma attenuation technique for measuring void fractions. Proc. of Conf. on application of high temperature instr. to liquid metal experiments. *ANL*-7100. [53]

Fincke, J R, Anderson, J L, Arave, A E, Deason, V A, Lassahn, G D, Goodrich, L B, Colson, J B and Fickas, E T (1978) Instrumentation for two-phase flow measurements in code verification experiments. *Paper presented at CNSI Specialist Meeting on Transient Two-Phase Flow, Paris, June, 1978.* [62,93,130]

Finlay, I C and Welsh, N (1968) A photographic technique for determining the velocities of liquid droplets flowing in an air-stream. *Jnl. Photogr. Sci.* **16**, 70-78. [152]

Fiori, M P and Bergles, A E (1966) A study of boiling-water flow regimes at low pressure. *Mass. Inst. Technol.* Report No. 5382-40. [107]

Firasat Ali, S (1975) Hot-wire anemometry in moderately heated flow, *Review of Scientific Instruments.* **46** (2), 185-191. [128]

Firstenberg, H (1960) Boiling songs and associated mechanical vibrations. *NDA*-2131-12. [76]

Fisenko, V V and Sychikov, V I (1977) Loss of pressure in a pipe with discharge of critical two-phase flow from it. *Thermal Eng.* **23** (8), 75-76. [72]

Fisher, S A and Yu, S K W (1975) Dryout in serpentine evaporations. *Int. J. Multiphase Flow.* **1** (6), 771-791. [203]

Fitremann, J M (1972) La debitmetrie electromagnetique appliquee aux emulsion. *Soc. Hydrotechnique de France Douziemes Journees de*

References

L'Hydraulique, Paris. Question IC Rapport 6. [63]

Fitzgerald, P D (1974) Laser-Doppler velocimeter evaluation and measurements in a flow with drag reduction. *Ph.D Thesis, Univ. Michigan, U. Microfilms*, Order No. 75-686. [135]

Flinta, J (1975) Blowdown of subcooled hot water. *European Two-Phase Flow Group Meeting, Haifa, 1975*. [72,73]

Flinta, J, Hernborg, G and Akesson, (1971) Results from the blowdown test for the exercise. *European Two-Phase Flow Group Meeting, RISØ, Denmark, June 1971*. [98]

Foglia, J J, Peter, F G, Epstein, H M, Wooton, R O, Dingee, D A and Chastain, J W (1961) Boiling-water void distribution and slip ratio in heated channels. *BMI* 1517. [49]

Fohrman, M J (1960 The effect of the liquid viscosity in two-phase two-component flow. *Argonne National Lab*. Report No ANL 6256. [49]

Foord, R, Harvey, A F, Jones, R, Pike, E R and Vaughan, J M (1974) A solid-state electro-optic phase modulator for laser Doppler anemometry. *Journal of Physics D: Applied Physics*. **7**, L36-L39. [138]

Forslund, R P and Rohsenow, W M (1966) Thermal non-equilibrium in dispersed-flow film boiling in a vertical tube. *Dept. Mech. Eng. MIT*. Report No 75312-44. [99]

Fouda, A E and Rhodes, E (1972) Two-phase annular flow stream division. *Trans. Inst. Chem. Eng.* **50**, 353. [89]

Fouda, A E and Rhodes, A E (1974) Two-phase annular flow stream division in a simple tee. *Trans. Inst. Chem. Eng.* **52**, 354. [89]

Fouda, A E and Rhodes, E (1976) Total mass flow and quality measurement in gas-liquid flow. *Paper presented at Two-Phase Flow and Heat Transfer Symp., Fort Lauderdale, Florida, October 1976*. [96]

Fourney, M Em Mattkin, J H and Waggoner, A P (1969) Aerosol size and velocity determination via holography. *Rev. Sci. Instrum.* **40**, 205-213. [152]

France, D M (1974) Liquid metal heated DNB experiments in high pressure forced convection boiling of Freon-12. *Symp. Multi-Phase Flow Systems. Univ Strathclyde, April 1974*. Paper No E4, (I. Chem. E. Symp. Series No 38). [68]

Franklin, D (1961) The measurement of rapid surface-temperature fluctuations during nucleate boiling of water. *AIChE Journal*, **7** (4), 620-624. [186]

Frantisak, F, Palade de Iribarne, A, Smith, J W and Hummel, R L (1969) Non-disturbing tracer technique for quantitative measurements in turbulent flow. *Ind. Eng. Chem. Fund.* **8** (1), 160-167. [183]

Freeze, P D and Parker, L P (1972) Bibliography of temperature measurement, Jan. 1953 to Dec. 1969. *National Bureau of Standards*, NBS special publication 373. [31,168]

Freymuth, P (1972) Improved linearisation for hot wire anemometers, *Journal of Physics E: Scientific Instruments*, **5**, 533-534. [128]

Friedel, L and Mayinger, F (1974) Scaling of two-phase friction pressure drop. *Paper presented at European Two Phase Flow Group Meeting, Harwell, June 1974*. Paper No B2. [203]

Friz, G and Riebold, W (1972) Experimental investigation of the post-crisis heat transfer in a cooling sub-channel of a PWR during depressurisation. *Crest Specialist Meeting on Emergency Core Cooling for Light Water Reactors, Munich*, MRR 115, **1**. [78]

Friz, G, Riebold, W, Lange, D and Megnin, J C (1975) Calculations compared with experiments simulating different blow-downs. *Trans. A.N.S.* **20**, 483. [72]

Fukuda, K and Kobori, T (1978) Two-phase flow instability in parallel channels. *Proceedings of the 6th International Heat Transfer Conference, Toronto, August 1978.* **1**, 369. [92]

Fuls, G M and Geiger, G E (1970) Effect of bubble stabilisation on pool-boiling heat transfer. *Jnl. Heat Transfer.* **92**, Series C (4), 635-640. [86]

Fultz, G F and Mesler, R B (1970) The measurement of surface temperatures with platinum films during nucleate boiling of water. *AIChE Journal.* **16**, 44-48. [188]

Gaertner, R F (1963) Photographic study of nucleate pool-boiling on a horizontal surface, *General Electric.* Report No 63-RL-3357C. [192]

Galaup, J P, Delhaye, J M (1976) Use of miniature optical probes in two-phase gas-liquid flow. Application to the measurement of the local velocity of the gas phase. *Houille Blanche*, **1**, 17-29. [

Galley, R (1974) (consulting ed.) Turbine flowmeters. *Measurements and data*, 102-109. [98]

Ganic, E N and Rohsenow, W M (1977) Dispersed flow heat transfer. *Int. J. Heat Mass Transfer*, **20** (8), 855-866. [39]

Garbarini, G R and Tien, C (1969) Mass transfer from single gas bubble: a comparative study on experimental methods. *Canad. Jnl. Chem. Engng.* **47**, 35-41. [42]

Gardiner, J A (1964) Measurement of the drop size distribution in water sprays by an electrical method. *Instrum. Practice*, **18**, 353-356. [153]

Gardner, R P, Bean, R H and Ferrell, J K (1970) On the gamma-ray one-shot-collimator measurement of two-phase flow void-fractions. *Nucl. Appl. and Techn.* **8**, 88-94. [49]

Gardon, R (1960) A transducer for the measurement of heat-flow rate. *Jnl. Heat Transfer.* **82**, 396-398. [38]

Garg, S C and Patten, T D (1965) Temperature and pressure transients near the heating surface during nucleate pool-boiling of saturated water. *Inst. Mech. Engrs.* 204-211. [186]

Gaspari, G P and Granzini, R (1976) Transient dryout model: comparison with data in a cirene 18+1 rod cluster. *European Two-Phase Flow Meeting, Erlangen, June 1976.* Paper No A10. [79,80]

Gaspari, C P, Granzini, R and Hassid, A (1971) Heat transfer crisis in transient conditions: experimental data for inlet flow-stoppage at constant pressure. *European Two Phase Flow Group Meeting, RISØ Denmark, 1971.* Paper D6. [79,80]

Gaspari, C P, Granzini, R and Hassid, A (1972) Dryout onset in flow stoppage depressurisation, and power surge transients. *Crest Specialist Meeting on Emergency Core Cooling for Light Water Reactors, Munich.* MRR 115, **1**. [79]

Gauvin, W H (1964) The motion of freely-moving single particles. *Conf. Dust Cooling, Queen Mary College, London.* 2, 34-48. [133]

Gavrilov, L R (1970) Free gas content of a liquid and acoustical techniques for its measurement. *Soviet Phy. Acoust.* **15**, 285-295. [62]

Geake, J E and Smalley, C (1975) Optical depth gauges and level

controllers for liquids. *Chem. Eng.* 301-304. [126]
Gerrard, J H (1971) An experimental investigation of pulsating turbulent water flow in a tube. *J. Fluid. Mech.* **46**, Pt 1, 43-64. [132]
Gibson, E J, Rennie, J and Say, B A (1957) The use of gamma radiation in the study of the expansion of gas-liquid systems. *Internat. J. Appl. Radiation and Isotopes*, **2**, 129-135. [49]
Gill, L E, Hewitt, G F and Hitchon, J W (1963a) Sampling-probe studies of the gas core in annular two-phase flow. Part 1. The effect of length on phase and velocity distribution. *AERE*-R3954 (1962) also, *Chem. Engng. Sci.* **18**, 525-535. [169]
Gill, L E, Hewitt, G F and Lacey P M C (1963b) Sampling-probe studies of the gas core in annular two-phase flow. Part 2. Studies of the effect of phase flow-rates on phase and velocity distriubtion. *AERE*-R3955. [112,169]
Gjessing, D T, Lanes, T and Tangerud, A (1969) A hot wire anemometer for the measurement of the three orthogonal components of wind velocity, and also directly the wind direction, employing no moving parts. *Journal of Scientific Instruments, Journal of Physics E*, **2**, Series 2, 51-54. [128]
Goldschmidt, V W (1965) Measurement of aerosol concentrations with a hot-wire anemometer. *Jnl. Colloid Sci.* **20**, 617-634. [125]
Goldschmidt, V and Eskinazi, S (1966) Two phase turbulent flow in a plane jet. *Jnl. Appl. Mech.* 735-747. [125]
Goldsmith, H L and Mas, S G (1963) The flow of suspensions through tubes. H. single large bubbles. *Jnl. Colloid. Sci.* **18**, 237-261. [184]
Golovin, V A, Konyaeva, N P and Rinkevichyos, B S (1971) Study of the model of a two-phase flow using an optical quantum generator (laser). *High Temp.* **9** (3), 549-553. [138]
Gomezplata, A and Brown, R W (1968) Axial dispersion coefficient measurement in two-phase flow. *AIChE Journal.* **14**, 657-658. [149]
Gonzalez-Santalo, J M (1971) Two phase flow mixng in rod bundle subchannels. *Ph.D Thesis MIT.* [149]
Gorman, D J, Cali, G P, Grillo, P and Testa, G (1970) Flow-induced vibrations. *Trans. Amer. Nucl. Soc.* **13** (1), 333-338. [86]
Gouse, S W (1964) Two-phase gas-liquid flow oscillations: preliminary survey. *Notes Special Summer Program. Two-Phase Gas-Liquid Flow, Mass. Inst. Techn.* [91]
Gouse, S W and Dickson, J A (1966) Heat transfer and fluid flows inside a horizontal tube evaporator, Phase 2. *ASHRA Trans.* [31, 106]
Gouse, S W and Hwang, C C (1963) Visual study of two-phase one-component flow in a vertical tube with a heat transfer. *MIT* Report No 8973-1. [26]
Graham, R W (1965) Experimental observations of transient boiling of subcooled water and alcohol on a horizontal surface. *NASA* TND-2507. [188]
Greated, C and Durrani, T S (1971) Signal analysis for laser velocimeter measurements, *Journal of Physics E: Scientific Instruments*, **4**, 24-26. [137]
Green, J R and Hauptmann, E G (1970) Forced convection heat transfer from a cylinder in carbon dioxide near the thermodynamic critical point. *ASME* 70-HT/Sp T-36. [35]

Green, S J (1967) Some experimental techniques used in reactor heat transfer and fluid flow research. *WAPD*-TM-386. [65]

Green, W and Mesler, R (1970) Experimental study of transient pressures accompanying vapour-bubble collapse in water. *ASME Symp. Role of Nucleation in Boiling and Cavitation, May 1970*. Detroit, Paper 70-FE-25. [77,192]

Greenkorn, R A, Matar, J E and Smith, R C (1967) Two-phase flow in Hele-Shaw models. *AIChE Journal*, **13**, 273-279. [196]

Gregory, G A and Scott, D S (1968) Physical and chemical mass-transfer in horizontal cocurrent gas-liquid slug flow. *Int. Symp. Res. In Cocurrent Gas-Liquid Flow, Waterloo, Ontario, Sept. 1968*. Paper F3, 2. [41]

Griffin, O M and Votaw, C W (1973) The use of aerosols for the visualisation of flow phenomena. *Int. J. Heat Mass Transfer*, **16**, 217-219. [130]

Griffith, P (1963) The slug-annular flow regime transition at elevated pressure. *Argonne Nat. Lab.* Report No. ANL-6796. [107]

Griffith, P (1978) Instrumentation problems concerning heat and mass transfer during LOCAS in nuclear reactors. *Lecture presented at Fluid Dynamics Institute, Short Course on Two-Phase Flow Measurements, Dartmouth College, New Hampshire, August 1978*. [69]

Griffith, P, Clark, J A and Rohsenow, W M (1958) Void volumes in subcooled boiling systems. *ASME* Paper No 58-HT-19. [63]

Griffith,, P, Kirchner, W and Smith, T (1975) Rewet heat transfer and carry-over. *Oral presentation for the AEC Reactor Safety Conf.* [82]

Grigoreiv, V A, Klimenko, B V, Pavlov, Y M and Ametistov, Y V (1978) The influence of some heating surface properties on the critical heat flux in cryogenic liquid boiling. *Proceedings of the 6th International Heat Transfer Conference, Toronto, August, 1978*. **1**, 215. [35]

Grolmes, M A and Fauske, H K (1969) Acoustic techniques for flow regime and gas detection. *ANS Trans*. **12** (2), 832-833. [76]

Grolmes, M A and Fauske, H K (1974) Axial propagation of free-surface boiling into superheated liquids in vertical tubes. *5th Internat. Ht. Transfer Conf. Sci. Council. Japan, Sept. 1974*. **4**. [77]

Gross, M, Jentges, H, Kafka, P and Stolben, H (1971) Experimental results on acoustic detection of sodium and water boiling. *Atomkernergie (ATKE)* Bd 17 Lfg 4. [76]

Gucker, F T and Egan, J J (1961) Measurement of the angular variation of light scattered from single aerosol droplets. *Jnl. of Colloid Science*, **16**, 68-84. [154]

Gucker, F T and Rowell, R L (1960) The angular variation of light scattered by single dioctyl phthalate aerosol droplets. *Disc. Faraday Soc.* **30**, 185-191. [154]

Gunkel, A, Patel, R P and Weber, M E (1971) A shielded hot-wire probe for highly turbulent flows and rapidly reversing flows. *Ind. Eng. Chem. Fund.* **10** (4), 627-631. [183]

Gunther, F C (1951) Photographic study of surface boiling heat transfer to water with forced convection. *Trans. ASME*, 115-123. [189]

Gustafsson, B, Harju, R and Imset, O (1974) Flow and enthalpy distribution in a 9-rod bundle. *Paper presented at European Two Phase Flow Group Meeting, Harwell, June 1974*. Paper No A4. [149]

Gustafsson, B and Kjellen, B (1971) Two-phase flow in a 9 rod bundle with inclined power distribution. *European Two-Phase Flow Group Meeting. RISØ Denmark, June 1971.* Paper A5. [49,66,122]

Haberstroh, R E and Griffith, P (1965) The slug-annular two-phase flow regime transition. *Amer. Soc. Mech. Eng.* 65 HT52. [107]

Hager, N E Jr. (1965) Thin-foil heat meter. *Rev. Scient. Instrum.* 36 (11), 1564-1570. [39]

Hahne, E and Diesselhorst, T (1978) Hydrodynamic and surface effects on the peak heat flux in pool boiling. *Proceedings of the 6th International Heat Transfer Conference, Toronto, August 1978.* 1, 209. [192]

Haigh, C P and Ponter, A B (1971) Sound emission from boiling on a submerged wire. *Canadian J. Chem. Engng.* 49 (3), 309-313. [76]

Hamilton, L J, Nyer, R and Schrock, V E (1967) Propagation of shock waves through two-phase, two-component media. *ANS Trans.* 10 (2). [71]

Hammitt, F C, Smith, W, Lauchlan, I E B, Iyany, R D and Robinson, M J (1964) Void fraction measurements in a cavitating mercury venturi. *Am. Nucl. Soc. Trans.* 7, No 1, 189-193. [49]

Hancox, W T, Forrest, C F and Harms, A R (1972) Void determination in two-phase systems employing neutron transmission. *ASME* Paper No 72-HT-2. [49]

Hanus, N, Gnadt, P A and Fontana, M H (1977) Local sodium boiling in a partially blocked simulated LMFBR subassembly (Thors bundle 3B) *ORNL* Report No TM-5862. [76]

Haque, H and Boothroyd, R G (1968-69) Heat and momentum transfer analogy in dense aerosols flowing turbulently in ducts. *Powder Technol.* 2, 305-307. [149]

Harms, A A and Forrest, C F (1971) Effects in radiation diagnosis of fluctuating voids. *Nucl. Sci. Engng.* 46, 408-413. [54,182]

Harms, A A and Laratta, F A R (1973) The dynamic bias interrogation of two-phase flow. *Int. J. Heat Mass Transfer,* 16, 1459. [53]

Harms, A A, Lo, S and Hancox, W T (1971) Measurement of time-averaged voids by neutron diagnosis. *Jnl. of Appl. Phys.* 42 (10), 4080-4082. [53]

Harris, D M (1967) Calibration of a steam-quality-meter for channel power measurement in the prototype SGHW reactor. *European Two-Phase Heat Transfer Meeting, Bournemouth, 1967.* [96,97]

Harris, J (1970) Interrogating flow field with radar and laser sources. *Inst. Meas. Conf. Symp. (Univ. Surrey), 1970.* [135]

Harris, D M and Shires, G L (1972) Two phase pressure drop in a venturi. *Nat. Engng. Lab. UK.* Report No 549, 18-33. [96]

Hassan, D, Orell, A and Fink, M (1970) Annular flow of two immiscible liquids L. Mechanisms. *Canad. Jnl. Chem. Engng.* 4, 514. [106

Hauptmann, G, Lee, V and McAdam, D (1973) Two-phase fluid modelling of the critical heat flux. *Proc. of International Meeting on Reactor Heat Transfer, Karlsruhe, October 1973.* 557. [206,207]

Hawes, R I (1976) Heater pins for LOCA simulations studies. *European Two-Phase Flow Meeting, Erlangen 1976.* Paper No N6. [81]

Hayashi, T (1974) Full-scale blow-down experiments for pressure-tube-type heavy water reactor. *Proc. JUICE Meeting, Fugen, Japan.* Report No 7. [75,82]

Hayes, D J (1969) Acoustic detection of gas bubbles in liquid sodium. *CEGB* RD/B/N1423. [76]

Hayes, D J (1972) A new method of continuously monitoring the void fraction of flowing sodium. *CEGB* RD/B/N2259. [63]

Haywood, R W, Knight, G A, Middleton, G E and Thom, J (1961) Experimental study of the flow conditions and pressure drop of steam/water mixtures at high pressure in heated and unheated tubes. *Proc. Instn. Mech. Engrs.* **175** (13), 669-749. [49]

Heckle, M (1970) Determination of two-phase gas/liquid flow with the aid of throttling devices. *Chem. Ing. Tech.* **42** (5), 304-310. [96]

Hein, D and Liebert, H (1976) The influence of pressure on the Leidenfrost temperature for Freon-12. *European Two-Phase Flow Meeting, Erlangen, June 1976.* Paper No D8. [83]

Heineman, J B, Marchaterre, J F and Mehta, S (1963) Electromagnetic flowmeters for void fraction measurement in two-phase metal flow. *Rev. Scient. Instr.* **34** (4), 399-401. [49]

Helmick, H H (1977) LASL de-entrainment instrumentation. *Paper presented NRC 5th Water Reactor Safety REs. Information Mtg. Gaithersburg, Maryland, USA, Nov. 1977.* [138]

Hendricks, R C and Sharp, R R (1964) Initiation of cooling due to bubble growth on a heating surface. *NASA* TN-D-2290. [188]

Hendrix, C D, Dave, S B and Johnson, H F (1967) Translation of continuous phase in the wakes of single rising drops. *AIChE Journal* **13**, 1072-1077. [192]

Henry, R (1975) Transient blowdown and CHF studies. Argonne Nat. Lab. *Third Water Reactor Safety Reserach Infor. Meeting, Washington DC Sept. 1975.* [72]

Henry, R E (1968) Study of one- and two-component two-phase critical flows at low qualities. *Argonne Nat. Lab.* Report No ANL-7430. [49]

Henry, R E and Leung, J C M (1977) Transient and steady-state critical heat flux experiments in Freon. *Paper presented NRC 5th Water Reactor Safety Res. Information Mtg. Gaithersburg, Maryland, USA. Nov 1977.* [79,80,203]

Heron, R A, Kimmeir, J H and Stevens, G F (1969) Burnout power and pressure drop measurements on 12 foot 7-rod clusters cooled by Freon-12 at 155 psia. *AEEW*-R655. [203]

Herringe, R A (1971) Measurement of voidage and slip distribution in two-phase flows with no interphase mass transfer. *Aus. Inst. Nucl. Sci. and Eng. 5th AINSE Heat Transfer and Fluid Flow Conference.* Paper 22. [122]

Herringe, R A and Davis, M R (1974) Detection of instanteous phase changes in gas-liquid mixtures. *J. Phys. E. Sci. Instrum.* **7** (10), 807-812. [122]

Herzberger, P and Bonvini, C (1973) The use of capacitance probes for transient void-fraction measurements in a two-phase flow water vapour. *Euratom Report* No EUR 5026. [57]

Herzberger, P and Hufschmidt, W (1974) Determination of the steam content and flow pattern of a two-phase mixture in concentric annular sondes. *Euratom Report* No EUR 5077. [57]

Herzfeld, C M et al (eds) (1962) "Temperature: its measurement and control in science and industry" **3**, Part 1 (Sub-ed. Brickwedde, F G) Part 2 (Sub-ed. Dahl, A I) Reinhold Publishing Corp, New York, [31]

References

Hetsroni, G, Cuttler, J M and Sokolov, M (1969) Measurements of velocity and droplet concentration in two-phase flows. *JAMCA*, 334-335. [184]

Hewitt, G F (1964) Photographic and entrainment studies in two-phase flow systems. *AERE*-R4683. [171,189]

Hewitt, G F (1966) Improvements in or relating to liquid film thickness measurements. *Brit. Patent Spec.* No 1051641. [180]

Hewitt, G F (1969-70) Disturbance waves in annular two-phase flow. *Proc. Instn. Mech. Engrs.* **184**, Pt 3C, 142-150. [177,180]

Hewitt, G F (1971) Role of experiments in two-phase systems with particular reference to measurement techniques. *Jnl. Brit. Nucl. Energy Soc.* **12** (2), 213-240. Also *Advances in Ht. and Mass Transfer.* 6. [Foreword]

Hewitt, G F (1977) Mechanism and prediction of burnout. *Lecture at NATO Advanced Study Institute, Istanbul, August 1976. Hemisphere Publishing Corp. Washington, 1977. Vol II,* 71. [13]

Hewitt, G F (1977) Two phase flow patterns and their relationship to two-phase heat transfer. *Lecture at NATO Advanced Study Institute, Istanbul, August 1976. Hemisphere Publishing Corp. Washington, 1977.* **I.** [5]

Hewitt, G F (1978) Critical heat flux in flow boiling. *Proceedings of the 6th International Heat Transfer Conference, Toronto, Aug. 1978.* **6**, 143. [65,151,207].

Hewitt, G F and Hall-Taylor, N S (1970) "Annular two-phase flow" Pergamon Press, Oxford. [1,10,11,13,64,105,112,169,171,184]

Hewitt, G F, Kearsey, H A, Lacey, P C M and Pulling, D J (1965) Burnout and film flow in the evaporation of water in tubes. *AERE*-R4864. [67,171]

Hewitt, G F, Kearsey, H A and Keeys, R K F (1969) Determination of rate of deposition of droplets in a heated tube with steam-water flow at 1000 psia. *AERE*-R6118. [150,151]

Hewitt, G F, King, I and Lovegrove, P C (1961) Holdup and pressure drop measurements in the two-phase annular flow of air-water mixtures. *AERE*-R3764. [61]

Hewitt, G F, King, R D and Lovegrove, P C (1962) Techniques for liquid film and pressure drop studies in an annular two-phase flow. *AERE*-R3921. [26,27,112,114,177]

Hewitt, G F and Lovegrove, P C (1962) The application of the light absorption technique to continuous film-thickness recording in annular two-phase flow. *AERE*-R3953. [177,178]

Hewitt, G F, Lovegrove, P C and Nicholls, B (1964) Film thickness measurement using a fluorescence technqiue. Part 1. Description of the method. *AERE*-R4478. [180]

Hewitt, G F and Lovegrove, P C (1969) A mirror-scanner velocimeter and its application to wave velocity measurement in annular two-phase flow. *AERE*-R3958.

Hewitt, G F and Nicholls, B (1969) Film thickness measurement in annular two-phase using a fluorescence spectrometer technique. Part 2: Studies of the shape of disturbance waves. *AERE*-R4506. [115,180]

Hewitt, G F and Roberts, D N (1969a) Studies of two-phase flow patterns by simultaneous X-ray and flash photography. *AERE*-M2159. [5,6,106]

Hewitt, G F and Roberts, D N (1969b) Investigation of interfacial phenomena in annular two-phase flow by means of the axial view technique. *AERE*-R6070. [190]

Hicken, E, Koch, E and Schad, O (1972) Heat transfer during blow-down with an inside cooling tube as test section. *Crest Specialist Meeting on Emergency Core Cooling for Light Water Reactors, Munich*. MRR 115, 1. [78]

Hills, R P T (1973) Annular two phase flow. An assessment of probe/tracer systems available for the study of mixing and entrainment phenomena. *CERL Laboratory* Memorandum No RD/L/M 427. [146]

Hinata, S (1972) A study on the measurement of the local void-fraction by the optical fibre-glass probe. *Bull. Japan. Soc. Mech. Engrs.* **15** (88), 1228-1235. [125]

Hoffer, M S and Resnick, W (1975) Electro-resistivity probe technique for steady- and unsteady-state measurements in fine dispersions. *Chem. Eng. Sci.* **30**, 473-502. [122,160]

Hoffmann de Visme, G F A and Singh, H (1972) Analysis of the thermal flowmeter operating from a pulsed-heat injection source. *J. Physics E. Scientific Instruments.* **5**, 885-888. [132]

Hoglund, B M, Weatherhead, R J and Epperson, T R (1961) Two-phase pressure drop in a natural circulation boiling channel. *Argonne National Lab.* Report No. ANL-5760. [57]

Hoggendoorn, C J (1959) Gas-liquid flow in horizontal pipes. *Chem. Engng. Sci.* **8**, 205-217. [57]

Hooker, H H and Popper, G F (1958) A gamma-ray attenuation method for void fraction determinations in experimental boiling heat transfer test facilities. *ANL*-5766. [49]

Hori, M, Kobori, T and Ouchi, Y (1966) Method for measuring void fraction by electromagnetic flowmeters. *JAERI* 1111. [63]

Hosler, E R (1965) Visual study of boiling water at high pressure. *AIChE Chem. Eng. Prog. Symp. Series*, **61** (57), 269-279. Also *WAPD*-T-1566. [189]

Hsiung, T H and Thodos, G (1977) Mass transfer in gas-fluidised beds: Measurement of actual driving forces. *Chem. Eng. Sci.* **32** (6). 581-592. [44]

Hsu, Y Y and Graham, R W (1963) A visual study of two-phase flow in a vertical tube with heat addition. *NASA* Tech. Note D-1564. [106,125]

Hsu, Y Y, Simon, F F and Graham, R W (1963) Application of hot-wire anemometry for two-phase flow measurements such as void fraction and slip velocity. *Multiphase Flow Symp. ASME, Philadelphia.* 26-34. [

Hsu, Y Y, Simoneau, R J, Simon, F F and Graham, R W (1969) Photographic and other optical techniques for studying two-phase flow. *Eleventh Nat. ASME/AIChE Ht. Transfer Conf. Minneapolis*, 1-23. [189]

Hu, CY and El-Wakil, M M (1974) Simultaneous heat and mass transfer in a rectangular cavity. *5th Int. Heat Transfer Conf. Tokyo, Sept. 1974.* **5**, Paper No CT1.5, 24-28. [142]

Hubbard, M G and Dukler, A E (1966) The characterisation of flow regimes for horizontal two-phase flow. 1. Statistical analysis of wall pressure fluctuations. *Heat Transfer Fluid. Mech. Conf.* 101-121. [108,185]

Hufschmidt, W and Burck, E (1972) Transient boiling heat transfer of

water at the vertical wall of a reactor pressure vessel. *Crest Specialist Meeting on Emergency Core Cooling for Light Water Reactors. Munich*, MRR 115, **1**. [80]
Hutchson, M W et al (1975) Experimental measurement of large pipe transient blowdown. *Trans. ANS.* **20**, 488. [93]
Huyghe, J and Minh, T G (1965) Flow of a mixture of gas and liquid in the dispersed annular region through a tube. Measurement of the amount of liquid flowing in the film on the wall. *CR Acad. Sci. Paris.* **260**, 2405-2408. [169]

Iida, Y and Kobayasi, K (1970) An experimental investigation on the mechanism of pool boiling phenomena by a probe method. *4th Internat. Heat Transfer Conf. Paris-Versailles.* **5**, Paper B1.3. [122]
Ilic, V (1974) Effect of pressure on burnout in annuli and a 19-rod cluster cooled by upflow of Freon-12. *Australian Atomic Energy Comm. Report* No AAEC/E324. [203]
Ilic, V (1974) The effect of pressure on burnout in a round tube cooled by Freon-12. *Australian Atomic Energy Comm.* Report No AAEC/E325. [203]
Imhoff, D H and Murray, J L (1972) Experimental basis for BWR emergency core cooling. *Crest Specialist Meeting on Emergency Core Cooling for Light Water Reactors, Munich*, MRR 115, **1**. [78]
Isbin, H S, Rodriguez, H A, Larson, H C and Pattie, B D (1959) Void fractions in two-phase flow. *AIChE Jnl.* **5** (4), 427-432. [49]
Isbin, H S, Sher, C and Eddy, K C (1957) Void fractions in two-phase steam-water flow. *AIChE Jnl.* **3** (1), 136-142. [49,50]
Ishigai, S, Yamane, M and Roko, K (1965) Measurement of component flows in a vertical two-phase flow by making use of the pressure fluctuations. *Bull. JSME.* **8** (51), 375-390. [96]
Ishii, M and Jones, O C (1976) Derivation and application of scaling criteria for two-phase flows. *Two-Phase Flow and Heat Transfer, NATO Advanced Study Inst. Istanbul, Turkey, Aug. 1976.* [205]
Iten, P D and Mastner, J (1974) A laser-Doppler velocimeter high spatial and temporal resolution. *In* "Flow - its measurement and control in science and industry", **1**, Part 2 (ISA) 1007-1013. [135]

Jackson, C (1976) A wheatstone bridge burnout detector. *AERE*-R8363. [65]
Jacobs, J D and Shade, A H (1969) Measurement of temperatures associated with bubbles in subcooled pool boiling. *Jnl. Ht. Trans.* 125-128. [193]
Jacowitz, L A and Brodkey, R S (1964) An analysis of geometry and pressure drop for the annular flow of gas-liquid systems. *Chem. Engng. Sci.* **19**, 261-274. [190]
Jagota, A K (1970) Hydrodynamics and mass transfer in upwards co-current gas-liquid annular flow in vertical tubes. *Ph.D Thesis. Univ. Waterloo.* [146]
Jagota, A K, Rhodes, E and Scott, D S (1973a) Tracer measurements in two-phase annular flow to obtain interchange and entrainment. *Canad. Jnl. Chem. Engng.* **51**, 139-148. [146]
Jagota, A K, Rhodes, E and Scott, D S (1973b) Measurement of resistance times and film and drop velocities in two-phase annular

flow. *Canadian J. Chem. Engng.* **51**, 393-400. [146]
Jagota, A K, Scott, D S and Rhodes, E (1972) Radial mixing and resistance time in gas-liquid annular flow in vertical tubes. *Canad. J. Chem. Eng.* 50, 194. [148]
Jagota, A K, Scott, D S and Rhodes, E (1973c) The measurement of liquid solute concentrations in gas-liquid dispersions. *Ind. Engng. Chem. Fundam.* 12, 131-140. [141]
James, K (1965-66) Metering of steam-water two-phase flow by sharp-edged orifices. *Proc. Instn. Mech. Engrs.* **180**, Pt 1, 549-571. [96]
James, L C (1965) Experiments on noise as an aid to reactor and plant operation. *Nucl. Engng.* 18-22. [76]
James, R (1961) Alternative methods of determining enthalpy and mass flow. *Proc. United Nations Co. Sources of Energy. Geothermal energy.* 2, 265-267. [99]
Janssen, E (1970) Two-phase flow and heat transfer in multirod geometries. *Gen. Elect. and AEC* Report No GEAP 10214. [107]
Janssen, E and Kervinen, J A (1964) Two-phase pressure drops in straight pipes and channels; water-steam mixtures at 600-1400 psia. *GEAP* 4616. [49]
Janssen, E, Kervinen, J A and Kim, H T (1971) Developing two-phase flow in tubes and annuli. *General Electric, US*. Report No GEAP-10341. [107,114]
Janssen, E and Schraub, F A (1968) Two-phase flow and heat transfer in multirod geometries. *GEAP*-5709. [117]
Jeandey, C and Pinet, B (1978) Experimental study of critical two-phase flow. *Paper presented at CNSI Specialist Meeting on Transient Two-Phase Flow, Paris, June 1978*. [54]
Jeffries, R B (1969) The structure of turbulence close to the interface of a co-current stratified two-phase flow. *Ph.D Thesis, Univ. Waterloo*. [184]
Jeffries, R B, Scott, D S and Rhodes, E (1970) Structure of turbulence close to the interface in the liquid phase of a co-current stratified two-phase flow. *Proc. I. Mech. E.* **184** (3C), 204-214 [183]
Jensen, A, Olsen, A and Mannov, G (1971) Film-flow experiments in annulus geometry. *European Two-Phase Flow Group Meeting, Risø, June 1971*. Paper A2. [114,171]
Jepsen, J C and Ralph, J L (1969-70) Hydrodynamic studies of two-phase upflow in vertical pipelines. *Proc. Instn. Mech. Engrs.* **184** (3C), 154-165. [118]
Johanns, J (1964) Development of a fluoroscope for studying two-phase flow patterns. *Argonne Nat. Lab.* ANL-6958. [106]
John, H, Muller, U and Reimann, J (1976) A test loop for testing measuring methods for the mass flow rate of two-phase flows. *European Two-Phase Flow Meeting, Erlangen, June 1976*. Paper No 84. [107,122]
Johnson, H A (1970) Transient boiling heat transfer. *4th Int. Heat Transfer Conf. Paris, 1970*. **5**, Paper B31. [78]
Johnson, H A and Abou-Sabe, A H (1952) Heat transfer and pressure drop for turbulent flow of air-water mixtures in a horizontal pipe. *Trans. ASME*, 977-987. [61]
Jones, A P (1977) A review of drop size measurement - the application

of techniques to dense fuel sprays. *Prog. Energy. Combust. Sci.* **3**, 225-234. [152]
Jones, O (1973) Statistical considerations in heterogeneous, two-phase flowing systems. *Private communication.* [54]
Jones, O C (1970) Determination of transient characteristics of an X-ray void-measurement system for use in studies of two-phase flow *Microfiche* KAPL 385P. [78]
Jones, O C (1977) BNL light water reactor thermohydraulic development programme; instrumentation tasks. *US Nuclear Regulatory Commission, Proceedings of Meeting of Review Group on Two-Phase Flow Instrumentation, January, 1977.* NUREG-0375. (Paper No I.3). [49, 135,136]
Jones, O C and Whitaker, S (1966) An experimental study of falling liquid films. *AIChE Journal,* **12**, 525-529. [178]
Jones, O C and Zuber, N (1974) Statistical methods for measurement and analysis in two-phase flow. *5th Internat. Ht. Transfer. Conf. Science Council, Japan.* **4**. [107,180,181]
Jones, O C and Zuber, N (1977) Interfacial passage frequency for two-phase, gas-liquid flows in narrow rectangular ducts. *Conf. on Heat and Fluid Flow in Water Reactor Safety, Manchester, Sep. 1977.* 5-10. [125]
Jordan, E L (1966) Instrumentation 3: Detection of in-core void formation by noise analysis. *ANS Trans.* **9** (1), 317-318. [76]
Judd, R L (1972) Comparison of experimental micro-layer thickness results. *Trans. CSME.* **1** (3), 168-170. [193]
Jurewicz, J T, Stock, D E and Crowe, C T (1977) Particle velocity measurements in an electrostatic precipitator with a laser velocimeter. *AIChE Symp. Series.* **73**, No 165, 138-141. [138]

Kaiser, A and Peppler, W (1977) Sodium boiling experiments in a seven-pin bundle: flow patterns and two-phase pressure drop. *Nucl. Eng. Des.* **43** (2), 285-293.
Kakac, S, Veziroglu, T N, Ergur, H S and Ucar, I (1978) The effect of inlet subcooling on sustained and transient boiling flow instabilities in a single channel up-flow system. *Proceedings of the 6th International Heat Transfer Conference, Toronto, August, 1978.* **1**, 363. [92]
Kamath, P S and Lahey, R T (1977) A turbine-meter evaluation model for two-phase transients (TEMT). *Report prepared for EG&G Idaho Inc. Renseleer Polytechnic Institute, October 1977.* Report NES-459. [98]
Kapur, D N and Macleod, N (1974) The determination of local mass transfer coefficients by holographic interferometry. *Int. J. Heat Mass Transfer,* **17**, 1151-1162. [44]
Kapur, D N and Macleod, N (1975a) Holographic determination of local mass transfer coefficients at a solid-liquid boundary. *AIChE Jnl.* **21** (1), 184-187. [44]
Kapur, D N and Macleod, N (1975b) The estimation of local heat transfer coefficients for 2-dimensional surface roughness elements by holographic interferometry. *Engng. Uses Coherent Optics Conf. Strathclyde,* 97-112. [57]
Karplus, H B (1961) Propagation of pressure waves in a mixture of water and steam. *ARF* 4132-12. [71]

Kartluke, H, Wichner, R P and Hoffman, H W (1965) An acoustic instrument for measuring sub-cooling in boiling system. *ORNL*-1678 Conf. 650946-3. [76]

Katarzhis, A K, Kosterin, S I and Sheinin, B I (1955) An electric method of recording the stratification of the steam-water mixture. *AERE* LIB/TRANSL. 590. [107]

Kawagde, M, Nadao, K and Otake, T (1975) Liquid-phase mass transfer coefficient and bubble size in gas sparged contactors. *J. Chem. Engng. Japan.* 8 (3), 254-256. [41]

Kawecki, W, Reith, T, van Heuven, J W and Beek, W J (1967) Bubble-size distribution in the impeller region of a stirred vessel. *Chem. Engng. Sci.* 22, 1519-1523. [152]

Kazemeini, M H and Ralph, J C (1975) Development of an ultrasonic technique for the measurement of single vapour-bubble dynamics in sodium. *AERE*-R7840. [77]

Kazin, I V (1964) Radial distribution of steam in a rising turbulent steam-water flow. *Teploenergetika,* 11, 40-43. [121]

Keeys, R K F, Ralph, J C and Roberts, D N (1970) The effect of heat flux on liquid entrainment in steam-water flow in a vertical tube at 1000 lb/sq in atm. (6.894 x 10^6 N/m^2). *AERE*-R6294. [171]

Kehler, P (1978) Two-phase flow measurement by pulsed neutron activation techniques. *Argonne National Laboratory,* Report ANL-NUREG-CT-78-17. [100]

Keller, A (1972) The influence of the cavitation nucleus spectrum on cavitation inception. Investigated with a scattered-light counting method. *J. Basic Engng.* 94, 917-925. [155]

Kemp, R F, Morse, A L, Wright, P W and Zivi, S M (1959) Kinetic studies of heterogeneous water reactors. *AEC Research and Dev.* Report RWD-RL-167. [51,61]

Kendron, J H, Stoner, E E and Taylor, G M (1963) Dynamic void-fraction measurement systems. *Atomics Int. (AEC)* Report No NAA-SR-7875. [51]

Keshock, E G (1974) A photographic study of flow condensation in 1-G and zero-gravity environments. *5th Internat. Ht. Transfer Conf. Tokyo.* Paper Cs 1.5, 236-240. [189]

Keyser, D R (1974) How accurate are thermocouples anyway? *Instruments and control systems.* 51-54. [32]

Kirchigin, A M (1970) Concerning certain possibilities of studying the boiling mechanism by acoustic methods. *Heat Transfer Soviet Res.* 2, 66-69. [76]

Kichigin, A M (1976) Sensitivity of the bridge and thermocouple methods for detection of boiling heat transfer crisis. *Heat Transfer Sov. Res.* 8 (3), 32-36. [65]

Kidron, I (1967) The signal-to-noise ratios of constant-current and constant-temperature hot-wire anemometers. *IEEE Trans. Instr. and Meas.* I-M-16, 68-73. [128]

Kielland, J B (1967) Propagation velocity of small amplitude pressure waves in steam-water mixtures. *Proc. Symp. on Two-Phase Flow Dynamics.* EUR-4288e. [71]

Kikuchi, V, Daigo, Y and Ohtsubo, A (1977) Local sodium boiling behind local flow blockage in simulated LMFBR fuel subassembly. *J. Nucl. Sci. Technol.* 14 (11), 774-790. [76]

Kikuchi, Y, Haga, K and Takahashi, T (1975) Experimental study of steady-state boiling of sodium flowing in a single-pin annular channel. *J. Nucl. Sci. Technol.* **12** (2), 89-91. [107]

Killian, W R and Simpson, J O (1959) Measuring vapour-liquid ratios during flow by a capacitance method. Advances in cryogenic engng. *Proc. of 1959 Cryo. Engng. Conf.* **5**, 505-508. [57]

Kinneir, J H, Heron, R A, Stevens, G F et al (1969) Burnout power and pressure drop measurements on 12ft 7 rod clusters coded by Freon-12 at 155 psia. *Presented at European Two-Phase Flow Meeting, Karlsruhe, June 1969, Paper V2.* Also *AEEW*-R655. [203]

Kinoshita, T and Murasaki, T (1969) On a stochastic approach to the pulsating phenomena in the two-phase flow. *Bull. JSME.* **12** (52), 819-826. [185]

Kirby, G J, Staniforth, R and Kinneir, J H (1965) A visual study of forced convection boiling, Part 1, results for a flat vertical heater. *AEEW*-R281. [189]

Kirchner, W L (1976) Reflood heat transfer in a light water reactor. *Massachusetts Inst. of Technol. Cambridge, Mass. USA*. Microfiche No NUREG-0106. **1-2**. [83]

Kirillov, P L, Smogalev, I P, Suvorov, M Y, Shumsky, R V and Stein, Y Y (1978) Investigation of steam-water flow characteristics at high pressures. *Proceedings of the 6th International Heat Transfer Conference, Toronto, August 1978.* **1**, 315. [114,118,161]

Kirkman, G A and Ryley, D J (1969) The use of laser photography for measuring the diameters of entrained droplets in two-phase flow. *Liverpool Univ. Dept. Mech. Engng. Report*. [152]

Kirsanov, A A, Markov, B L and Frolov, V N (1973) Thermal inserts for measuring the temperature of waterfall tubes and heat flux at their surfaces. *Thermal Engng.* **20** (9), 99-100. [38]

Kitayama, Y (1972) Digital void velocimeter. *J. Nucl. Sci. Technol.* **9** (10), 613-617.

Klimenko Yu, G, Rabinovich, M I and Skatynskaya, V Ye (1970a) Heat transfer from a cylindrical wall to a spouting bed. *Heat Transfer Soviet Research*, **2** (1). [38]

Kobayasi, K (1974) Measuring method of local phase velocities and void fraction in bubbly and slug flows. *5th Int. Heat Transfer Conf. Presentation at Round Table Discussion on Momentum and Heat Transfer, Sept. 1974.* [122,133]

Kokernak, R P and Feldman, C L (1972) Velocity of sound in two-phase flow of R12. *ASHRAE Journal*, **14** (2), 35-38. [71]

Kolb, B (1962) Measuring the size of droplets in wet steam. *Brown Boveri Rev.* **49**, 350-359. [153]

Kolbel, H, Beinhauer, R and Langemann, H (1972) Dynamic measurement of the relative gas content in bubble columns by X-ray absorption. *Chemie-Ing-Techn.* 44 (11), 697-704. [182]

Koloini, T, Sopcic, M and Zumer, M (1977) Mass transfer in liquid-fluidised beds at low Reynolds numbers. *Chem. Eng. Sci.* **32** (6), 637-641. [44]

Komasawa, I, Kuboi, R and Otake, T (1974) Fluid and particle motion in turbulent dispersion. 1. Measurement of turbulence of liquid by continual pursuit of tracer particle motion. *Chem. Engng. Sci.* **29** (3), 641-650. [131]

Kondic, N N and Hahn, O J (1970) Theory and application of the parallel and diverging radiation beam method in two-phase systems. *4th Int. Heat Transfer Conf. Paris, 1970)*, **7**, Paper MT 1.5. [53,55]

Kondic, M N and Lassahn, G D (1978) Nonintrusive density distribution measurement in dynamic high temperature systems. *Proceedings of the 24th International Instrumentation Symposium, 1978.* [55]

Konobeev, B I, Malyusov, V A and Zharvoronkov, N M (1957) Mass transfer in thin films of liquids. *Proc. Acad. Sci. USSR, Chem. Tech. Section,* 91-95. [42,112]

Kordyban, E (1977) The transition to slug flow the presence of large waves. *Int. J. Multiphase Flow,* **3**, 603. [106]

Kordyban, E S and Ranov, T (1963) Experimental study of the mechanism of two-phase flow in horizontal tubes. *Multiphase Flow Symp. ASME Philadelphia, Nov 1963.* [189]

Korsunskii, M I et al (1960) The use of radioactive isotopes for the measurement of moisture in steam. *Izmeritel'naya Tekh.* (5), 50-52. [54]

Kosky, P G and Henwood, G A (1968) A new technique for the experimental investigation of the collapse of vapour-filled bubbles in subcooled liquids. *AERE*-R5795. [192]

Kostic, L (1976) Local steam velocity measurements in a BWR using reactor noise analysis techniques. *Atomwirtschaft,* **21** (7), 363-367.

Kovacs, A and Mesler, R B (1964) Making and testing small surface thermocouples for fast response. *Rev. Scient. Instrum.* 35 (4) [188]

Kowalczewski, J J (1966) Vapour slip in two-phase fluid flow. *Trans. Am. Soc. Heat Refng. Air Condit. Engrs.* **72**, Pt 1. [121]

Kozeki, (1973) Film thickness and flow boiling for two-phase annular flow in helically coiled tubes. *Proc. Int. Meeting, Reactor Heat Transfer, Karlsruhe, October 1973.* [114]

Kraszewski, A (1971) Determination of water content in bi-phase amorphous mixtures by micro-wave method. *Proc. 1971, Eur. Microwave Conf.* **2**, CB/2.1-2.4. [63]

Kremnev, O A, Satanovskiy, A L, Protsyshin, B N and Tarasyuk, E N (1970) The interaction of water spheroids with heated surfaces. *Heat Transfer Soviet Research.* **2** (1). [83]

Kress, T S (1972) Mass transfer between small bubbles and liquids in cocurrent turbulent pipeline flow. *Microfiche* No ORNL-TM-3718. [41]

Kubie, J and Gardner, G C (1977) Flow of two liquids in a helix: An analogue of high pressure helical boilers. *Int. J. Multiphase Flow* **3** (4), 353-366. [196]

Kudo, A (1974) Basic study on vapour suppression. *5th Internat. Ht. Transfer Conf. Tokyo, Sept. 1974.* Paper CS 1.2, 221-225. [77]

Kulic, E and Rhodes, E (1977) Direct contact condensation from airstream mixtures on a single droplet.*Can. J. Chem. Eng.* 55 (2), 131-137. [43]

Kulic, E, Rhodes, E and Sullivan, G (1975) Heat transfer rate predictions in condensation on droplets from air-steam mixtures. *Canadian J. Chem. Engng.* **53** (3), 252-258. [43]

Kunz, H (1977) Selected topics in radiation thermometry. *Kerntechnische Gesellschaft im Deutschen Atomforum ev.* Proceedings of a Specialist Conference, February-March 1977, Hannover; "Experimental techniques in the field of thermodyanmics and fluid dynamics. Part II; thermometry techniques and simulation of thermodynamics processes. 55, Interatom, Berg-Gladbach, West Germany. [32,34]

Kutateladze, S S (1973) "Study of turbulent two-phase flow". Inst. Heat Phys. Siberian Branch Academy Sci. USSR, Novosibirsk. [22,

107,165,194]

Kutateldaze, S S, Burdukov, A P, Nakoryakov, V Ye, and Kuzmin, V Z (1969) Application of an electro-chemical method for measurement of shear stress in two-phase flow. *Heat Transfer Soviet. Res.* **1**, 66-73. [165]

Kutateladze, S S, Nakoryakov, V E and Burdukov, A P (1972) Spectral density of fluctuations of friction in a turbulent wall flow. *Sov. Phys. Doklady*, **16** (2), 87-89. [165,194]

Kutateladze, S S, Nakoryakov, V E and Burdukov, A P (1972) Spectral characteristics of vertical two-phase flow. *Soviet Phys. - Doklady*, **16**, 718. [165,194,195]

Kutukcuoglu, A, Perren, B and Varadi, G (1967) A method of detecting the onset of subcooling boiling by means of a vibration transducer. *Symp. Two-Phase Flow Dynamics, Eindhoven*. Session 11, Paper 2. [76,77]

Kutz, M (1968) "Temperature control" John Wiley and Sons Inc. New York, 108-127.

Lackme, C (1964) Analyse statistique de la structure locale d'un ecoulement diphase. I-Description des arrivees de bulles. *CENG*, Not TT, No 162. [122]

Lackme, C (1965) Some statistical properties of two-phase flow in vertical tubes. *Symp. on Two-Phase Flow, Exeter. 1965*. Paper D2. [122]

Lacke, C (1967) Structure and kinematics of two-phase bubble flow. *CEA* 3203. [122,133]

Lading, L (1971) Two-phase flow measurements utilizing laser anemometry. *Danish Atomic Energy Comm.* Report No RISØ-M-1369. [138]

Lafferty, R H (1971) Isotope technology development. Flow measurements with radio-isotopes. *Isotopes and Radiation Technol.* **8** (3), 287-309. [133]

Lafferty, J F and Hammitt, F G (1967) A conductivity probe for measuring local void fractions in two-phase flow. *Nucl. Applic.* **3**, 317-323. [127]

Lahey, R T (1977) USNRC sponsored instrumentation research at Rensselaer Polytechnic Institute (RPI), *US Nuclear Regulatory Commission, Proceedings of Meeting of Review Group on Two-Phase Flow Instrumentation, January 1977*. NUREG-0375, (Paper No I.2)

Lahey, R T (1978) A review of selected void fraction and phase velocity measurement techniques. *Lecture presented at Fluid Dynamics Institute Short Course on Two-Phase Flow Measurements, Dartmouth College, New Hampshire, August, 1978*. [47,100]

Lahey, R T and Moody, F J (1977) "The thermal hydraulics of a boiling water nuclear reactor" Monograph in the Nuclear Science Technology Series of the American Nuclear Society, Illinois, USA. [19]

Lahey, R T, Shiralkar, B S, Radcliffe, D W et al (1971) Out-of pile subchannel measurements in a nine-rod bundle for water at 1000 psia. *Int. Symp. Two-Phase Systems, Haifa, Israel, Sept 1971, Progress in heat and mass transfer.* **6**, 345. [149]

Landa, I and Tebay, E S (1972) The measurement and instantaneous display of bubble-size distribution, using scattered light, *IEEE Trans. Instrum. Measurements.* **21** (1), 56-59. [155]

Landa, I, Tebay, E S, Johnson, V and Lawrence, J (1970) The measurement of bubble-size distribution, using scattered light. *Hydraulics Inc.* Report TR 707-4. [155]

Landau, J, Boyle, J, Gomaa, H G et al (1977) Comparison of methods for measuring interfacial areas in gas-liquid dispersions. *Can. J. Chem. Eng.* **55** (1), 13-18. [152]

Landram, C S (1974) Transient-flow heat transfer measurements using the thin-skin method. *J. Heat Transfer*, **96** (3), 425-426. [39]

Langdon, W R, Bennett, W K, Decker, W T and Garland, W E (1970) Radiation effects on piezo-resistive accelerometers. *IEEE Trans. Indus. Elect. and Control Instr.* IECI-17 (2), 99-104. [76]

Langlois, G E, Gullberg, J E and Vermeulen, T (1954) Determination of interfacial area in unstable emulsions by light transmission. *Rev. Sci. Instrum.* **25**, 360-363. [154]

Larsen, D G (1975) Loft liquid-level probe tests in FLECHT-SETS. *Nuclear Techn. Div. Ann. Prog. Report for Period ending June 30 1974*. ANCR-1177, 238-239. [103]

Lassahn, G D (1975a) Mathematical analysis of a two-beam gamma densitometer. *Aerojet Nuclear Company*. (UC - 78) ANCR-1201. [50, 109,121]

Lassahn, G D (1975b) Two-phase flow velocity measurement with radiation intensity correlation. *Aerojet Nuclear Company*. (NRC-2) ANCR-1216. [139]

Lassahn, G D (1976) Two-phase flow velocity measurement using radiation intensity correlation. *ISA Transactions*. **15** (4), 297-300. [139]

Lassahn, G D (1977) LOFT three-beam densitomer data interpretation. *TREE*-NUREG-1111. [109,121]

Lassahn, G D and Arave, A E (1974) Drag disk, turbine, single-phase flow fret tests. *Nucl. Techn. Div. Ann. Progr. Report for period ending June 30 1974*. ANCR-1177, 250-251. [98]

Lavalee, H C and Popovich, A T (1974) Fluid flow near roughness elements investigated by photolysis method. *Chem. Eng'g Prog.* 29, 49-59. [162]

Laverty, W F and Rohsenow, W M (1967) Film boiling of saturated nitrogen flowing in a vertical tube. *Trans. ASME*, 90-98. [35]

Lawford, V N (1974) Differential-pressure transducers: the universal instrument for process control. *Advances in Instrumentation*. 29, Pt 4. 811 (1-13) ISA. [28]

Lecroart, H and Lewi, J (1972) Local measurements and their statistical interpretation for a high-velocity and high-vacuum level two-phase flow. *CEA*-CONF-2043. [122,133]

Lecroart, H and Porte, R (1971) Electrical probes for study of two-phase flow at high velocity. *Int. Symp. Two-Phase Systems, Haifa, Israel, Aug-Sept 1971, Session 3*. Paper 11. [133]

Lee, S L and Einav, S (1971) Migration in a laminar suspension boundary layer measured by the use of two-dimensional laser-Doppler anemometer. *Int. Symp. Two-Phase Systems, Israel*. Paper 3-14. [138]

Lee, S L, Srinivasan, J (1977) Development of laser-Doppler-anemometer technique to study droplet hydrodynamics in LOCA reflood. *Paper presented NRC 5th Water Reactor Safety Res. Information Mtg. Gaithersburg, Maryland, USA, Nov 1977*. [138,156,185]

Lee, S L and Srinivasan, J (1978) An experimental investigation of dilute two-phase dispersed flow using LDA technique. *Proceedings of 1978 Heat Transfer and Fluid Mechanics Institute, Stanford Univ. Press.* [138,156,157,158]

Lee, S L and Srinivasan, J (1978) Laser-Doppler anemometry technique applied to two-phase dispersed flows in a rectangular channel. *Proceedings of 1978 Two Phase Flow Instrumentation Review Group Meeting, Troy, New York, March 1978.* [138]

Lee, S L and Srinivasan, J (1978) Measurement of local size and velocity probability density distributions in two-phase suspension flows by laser-Doppler technique. *Int. J. Multiphase Flow*, **4**, 141. [138]

Lee, Y J (1978) An application of holography to the study of air-water, two-phase critical flow. *Ph.D. Thesis. Univ. Washington, 1973.* U. Microfilms order no 73-22. 586. [132,152]

Lee, Y, Chen, W J and Groeneveld, D C (1978) Rewetting of very hot vertical and horizontal channels by flooding. *Proceedings of the 6th International Heat Transfer Conference, Toronto, August 1978.* **5**, 95. [82]

Lee, Y J, Fourney, M E and Moulton, R W (1974) Determination of slip ratios in air-water two-phase critical flow at high quality levels utilising holographic techniques. *AIChE Journal*, **20** (2), 209-219. [132,152]

Legendre, P J (1974) Parameters affecting the accuracy of isothermal thermocouples, *5th Internat. Ht. Transfer Conf. Japan. Sept. 1974* Paper MA2.6 [167]

Lehmann, B (1968) An optical method for the measurement of local particle velocities in two-phase flows. *Symp. Electrical Power Generation, Warsaw, 1968.* Paper SM-107/8. [138]

Lehmkuhl, G D (1972) Time-domain reflectometry for liquid-level measurement. *Dow Chemical, USA*, AEC RFP-1902 (UC-37 Instruments) TID-4500. [103]

Leonchik, B I, Danilon, O L and Tynybekov, E K (1970) Rapid method for determining the dispersity of droplets of pure liquids. *Thermal Engng.* **17** (6), 101-103. [160]

Leung, J et al (1975) Critical heat flux during blow-down with reversed flow. *Trans. ANS.* **20**, 486. [80]

Leung, J C M (1976) Occurrence of critical heat flux during blowdown with flow reversal. *Ph.D Thesis.* ANL/RAS/LWR 76-1. [203]

LeVert, F E and Helminski, E (1973) A dual-energy method for measuring void fractions in flowing mediums. *Nuclear Technology*, **19**, 58-60. [49,53]

Lewi, J (1973) Local structure of gas-liquid bubble flow. *45th Euromech Colloquium, Palermo. Oct 1973.* [122,133]

Lilleleht, L U and Hanratty, T J (1961) Measurement of interfacial structure for cocurrent air-water flow. *Jnl. Fluid Mech.* **2**, 65-81. [178,194]

Lin, C S, Moulton, R W and Putnam, G L (1953) Interferometric measurements of concentration profiles in turbulent and streamline flow. *Ind. Eng'g Chem.* **45** (3), 640-646. [141]

Lindstrom, K, Kjellander, H and Jonsson, C (1970) A new instrument for the measurement of liquid level. *Review Sci. Instr.* **41** (7), 1085-1087. [103]

Lockett, M J and Kirkpatrick, R D (1975) Ideal bubbly flow and actual flow in bubble columns. *Trans. Instn. Chem. Engrs.* **53** (4), 267-273. [49]

Lockett, M J and Safekourdi, A A (1977) Light transmission through bubble swarms. *AIChE J.* **23** (3), 395-398. [154]

Lockhart, R W and Martinelli, R C (1949) Proposed correlation of data for isothermal two-phase two-component flow in pipes. *Chem. Eng. Prog.* **45**, 39. [11]

Lohmann, A W and Shuman, C A (1973) Image holography through convective fog. *Optics Communications* **7** (2), 93-97. [152]

Longwell, J P and Weiss, M A (1953) Mixing and distribution of liquids in high-velocity air-streams. *Industr. Engng. Chem.* **45** (3), 667-677. [117,150]

Lord, R G (1974) Hot-wire probe end-loss corrections in low density flows. *J. Physics E. Scientific Instruments*, **7**, 56-60. [128]

Lorenz, J J and Ginsberg, T (1973) Results and analysis of 7-pin wire-wrap mixing experiments. *Amer. Nucl. Soc. Trans.* **15** (2), 867-868. [149]

Lorenzi, A and Pisoni, C (1973) Experimental study of the stirred liquid motion in two-phase, two-component, cocurrent fluids. *La Termotecnia*, **27** (7), 350-354. [135]

Lottes, P A (1967) Shaped collimators improve one-shot void detectors. *PWR Reactor Technol.* **10** (2), 148-149. [49,53]

Lowe, P A (1975) ECC and steam interaction in the cold-leg piping. *Trans. ANS*, **20**, 497-501. [83]

Ludewig, E H W (1972) Photoelectric method of measuring size distribution of moderately dispersed drops in an immiscible binary liquid mixture. *Chemie-Ing-Techn.* 44 (5). [160]

Lurie, H and Johnson, H A (1962) Transient pool-boiling of water on a vertical surface with a step in heat generation. *Jnl. of Heat Transfer*. 217-224. [77]

Lyall, E (1969) The photography of bubbling fluidised beds. *Brit. Chem. Engng.* **14**, 501-506. [189]

Lykov, E V (1972) Thermoacoustic effects in surface boiling liquids. *Int. J. Heat Mass Transfer*. **15**, 1603-1614. [76]

Lynch, G F and Segel, S L (1977) Direct measurement of the void fraction of a two-phase fluid by nuclear magnetic resonance. *Int. J. Heat Mass Transfer*, 20 (1), 7-14. [63]

Lynnworth, L C (1975) Clamp-on ultrasonic flowmeters, *Instrumentation Technology*, 37-44. [128]

Lynnworth, L C, Pedersen, NE and Bradshaw, J E (1976) Ultrasonic flow meter optimisation study for two-phase prepolymer. *IEEE Group on sonics and ultrasonis, 1976 Ultrasonics Symposium, Annapolis, USA, Sept-Oct 1976.* [93]

Macleod, I D (1966) Some measurements of the acoustic spectrum produced by sub-cooled nucleate boiling. *UKAEA Report No TRG 1205 (R).* [76]

Macvean, S S and Wallis, G B (1969) Experience with Wicks-Dukler probe for measuring drop size distribution in sprays. *Dartmouth College Report.* [153]

Maddock, C, Lacey, P M C and Patrick, M A (1974) The structure of two-phase flow in a curved pipe. *Inst. Chem. Engrs. Symp.* No 38,

2, Paper No J2. [114,172]
Madsen, N (1965-66) Temperature fluctuations at a heated surface supporting pool-boiling water. *Proc. Instn. Mech. Engrs.* **180**, Pt 3C, 150-159. [188]
Magiros, P G and Dukler, A E (1961) Entrainment and pressure drop in co-current gas-liquid flow: 2 liquid property and momentum effects. *Devel. Mechanics*, **1**, 532-553. [169]
Magrini, U (1966) Flowmeter for measurement of low-velocities in liquids with weak electrical conductance. *Rev. Scient. Instrum.* **37** (5), 627-631. [133]
Mah, C C and Golding, J A (1971) Effect of flow pattern on gas-phase-controlled mass transfer. *Canad. Jnl. Chem. Engng.* 49, 160-162. [41]
Maitra, D and Raju, K S (1975) Vapour void-fraction in subcooled flow boiling. *Nucl. Engng. Design*, **32** (1), 20-29. [57]
Malnes, D (1966) Slip ratios and friction factors in the bubble flow regime in vertical tubes. *Kjeller*, KR-110. [122]
Mandhane, J M, Gregory, G A and Aziz, K (1974) A flow pattern map for gas-liquid flow in horizontal pipes. *Int. J. Multiphase Flow*, **1**, 537. [5,6]
Marcus, B D and Dropkin, D (1965) Measured temperature profiles within the superheated boundary layer above a horizontal surface in saturated nucelate pool-boiling of water. *Jnl. Heat Transfer.* **87**, 333-341. [188]
Marchaterre, J F and Petrick, M (1956) The effect of pressure on boiling density in multiple rectangular channels. *ANL*-5522. [49]
Marinelli, V and Pastori, L (1972) Experimental investigation of mass velocity distribution and velocity profiles in an LWR rod bundle. *European Two-Phase Flow Group Meeting, CNEN Rome*. Report No PR-PU (72) 7, Paper No C9. [149]
Marshall, J and Holland, P G (1977) Blowdown into full-pressure containment. *Conf. on Heat and Fluid Flow in Water Reactor Safety, Manchester, Sept. 1977*. 119-123. [72]
Martin, B W and Sims, G E (1971) Forced convection heat-transfer to water with air injection in a rectangular duct. *Int. J. Heat Mass Transfer.* 14, 1115-1134. [189]
Martin, G E and Grohse, E W (1961) X-ray absorption measurement of steady voids in water at high pressures. *TID*-14156. [51]
Martin, R (1969) Measurement of the local void fraction at high pressures in a heating channel. *Commissariat a l'Energie Atomique*, Report No CEA-R-3781. [51,121]
Martin, R (1972) Measurements of the local void fraction at high pressure in a heating channel. *CENG* BP 269-38. Grenoble. [51,52]
Martini, R and Premoli, A (1972) Bottom flooding experiments with simple geometries under different EEC conditions. *Crest Specialist Meeting on Emergency Core Cooling for Light Water Reactors, Munich*, MRR 115, **1**. [82]
Marto, P J and Sowersby, R L (1970) A photographic investigation of bubble nucleation from glass cavities. *ASME Symp. Role of Nucleation in Boiling and Cavitation, Detroit*, Paper 70-HT-16. [192]
Marucci, G, Nicodemo, L and Acierno, D (1968) Bubble coalescence under controlled conditions. *Int. Symp. on Res. Cocurrent Gas-Liquid Flow. Univ. of Waterloo. Canada, Sept 1968*. [192]

Masilyah, J H and Nguyen, T T (1974) Qualitative study in mass transfer by laser holography. *Canad. J. Chem. Engng.* **52** (5), 664-665. [142]

Matekunas, F A and Winter, E R F (1971) An interferometric study of nucleate boiling. *Int. Symp. Two-Phase Systems, Haifa, Israel, Aug-Sept 1971.* Session 1, Paper 15. [188]

Matkin, J H (1968) Determination of aerosol size and velocity by holography and steam-water critical flow. *Dissertation abstracts.* **29** (B), 2411. [132,152]

Matthes, W, Riebold, W and de Cooman, E (1970) Measurement of the velocity of gas bubbles in water by correlation method. *Rev. Sci. Instrum.* **41** (6), 843-845. [139]

Mattson, R J and Hammitt, F G (1972) A photographic study of subcooled flow boiling and the boiling crisis in Freon-113. *Microfiche* No PB 207 046. [189]

Mattson, R J, Hammitt, F G and Tong, L S (1973) A photographic study of the subcooled flow boiling crisis in Freon-113. *ASME* Paper No 73-HT-39. [189]

Maxwell, J (1881) "A treatise on electricity and magnetism" Clarendon Press, Oxford. [57]

Mayinger, F (1977) Simularity and model approaches to gas-liquid flows. *Kerntechnische Gesellschaft im Deutschen Atomforum ev*, Proceedings of a Specialist Conference, Feb-March 1977, Hannover; "Experimental techniques in the field of thermodynamics and fluid dynamics. Part II; thermometry techniques and simulation of thermodynamics processes", 71, Interatom, Berg-Gladbach, West Germany. [204]

Mayinger, F and Langner, H (1977) Use of an optical measurement technique to determine the entrainment behaviour of a two-phase flow in steady-state and non-stationary blow down conditions. *Kerntechnische Gesellschaft im Deutschen Atomforum ev.* Proceedings of a Specialist Conference, Feb-March 1977, Hannover; "Experimental techniques in the field of thermodynamics and fluid dynamics. Part II; thermometry techniques and simulation of thermodynamics processes", 42, Interatom, Berg-Gladbach, West Germany. [191]

Mayinger, F, Langner, H and Seiffert, V (1977) Experimental and theoretical investigations in reactor fluid behaviour. *European Two-Phase Flow Group Mtg. Grenoble, June 1977, Paper No B7.* [149]

Mayinger, F, Langner, H and Zetzmann, K (1976) Flow pattern of two-phase flow in inside cooled tubes - a generalised form of flow pattern map and entrainment measurements of annular flow. *European Two-Phase Flow Meeting, Erlangen, June 1976.* Paper No D9. [106,152,169]

Mayinger, F, Nordmann, D and Panknin, N (1974) Holographic investigation of subcooled boiling. *Chemie. Ing. Techn.* **46** (5). [193]

Mayinger, F and Panknin, W (1974) Holography in heat and mass transfer. *5th International Heat Trans. Conf. Sci. Council, Japan. Sept 1974.* Paper 1L3. [142,143,168,188,193]

Mayinger, F, Schad, O and Weiss, O (1967) Investigation into the critical heat-flux to boiling water. *Euratom Report* No EUR 3347e 1811. [66]

Mayinger, F, Zetzmann, K (1976) Flow pattern of two-phase flow in inside cooled tubes - a generalised form of flow pattern map, based on investigations in water and Freon. *Two Phase Flow and Heat Transfer*, NATO Advanced Study Inst. Istanbul, Turkey. Aug 1976. [203]

Mayinger, F and Zetzmann, K (1977) Vapour-content measurement, using the X-ray flash technique. *Kerntechnische Gesellschaft im Deutschen Atomforum ev.* Proceedings of a Specialist Conference, Feb-March 1977, Hannover; "Experimental techniques in the field of thermodynamics and fluid dynamics. Part II; thermometry techniques and simulation of thermodynamics processes" 415. Interatom, Berg-Gladbach, West Germany. [106]

Mayo, W T Jr. (1970) Simplified laser-Doppler velocimeter optics. *Journal of Physics E: Scientific Instruments*, **3**, 235-237. [137]

Mazzone, G (1974) Fluid redistribution studies in two-channel test sections. *CNEN* Report No RT/ING(74)14. [149]

McCalvey, L F and Thompson, A (1973) Development of special transducers for vibration measurements in hostile environments. *Int. Symp. Vibration Problems in Industry, Keswick. April 1973*. Paper No 626. [86]

McLeod, W R, Rhodes, D F and Day, J J (1971) Radiotracers in gas-liquid transportation problems - a field case. *J. Petrol. Technol.* **23**, 939-947. [99]

McManus, H N (1957) An experimental investigation of film characteristics in horizontal annular two-phase flow. *ASME* Paper 57-A-144. [26]

McPherson, G D (1971) The use of Freon to model dryout in a high pressure water system. *Atomic Energy Canada Ltd.* Report No AECL-3787.

McPherson, G D (1977) Results of the first three non-nuclear tests in the LOFT facility. *Nucl. Safety.* **18** (3), 306-316. [73,93]

McShane, J L (1974) Ultrasonic flowmeters *In* "Flow - its measurement and control in science and industry", **1**, Part 2 (ISA), 897-915. [128]

Mecredy, R C, Wigdortz, J M and Hamilton, L J (1970) Prediction and measurement of acoustic wave propagation in two-phase media. *ANST* **13**, Part 2, 672. [71]

Mellen, R H (1954) Ultrasonic spectrum of cavitation noise in water. *Journ. Acoust. Soc. Amer.* **26** (3), 356-360. [77]

Mensing, W and Schugerl, K (1970) Mass transfer measurements on suspended droplets (Part 1: Measuring technique). *Chem. Ing. Tech.* **42** (12), 837-841. [44]

Merilo, M (1977) Critical heat flux experiments in a vertical and horizontal tube with both Freon-12 and water as coolant. *Nucl. Eng. Des.* **44** (1), 1-16. [203]

Merilo, M, Dechene, R L and Cichowlas, W M (1977) Void fraction measurement with a rotating electric field conductance gauge. *J. Heat Transfer*, **99**, 330. [57,59,60,206,207]

Mesler, R (1975) Enhanced cooling due to bubble merger during nucleate boiling. *Paper presented at 79th Nat. AIChE Meeting, Houston, March 1975.* [188]

Michiyoshi, I and Nakajima, T (1964) An experimental study of bubble motion on a heating surface in nucleate boiling. *Mem. Fac. Engng. Kyoto Univ. Japan*, **26**, 336-352. [192]

Michiyoshi, I, Ohtsubo, A and Takayasu, M (1969) Device for measuring local heat flux on the surface of a fuel element. *Nucl. Engng. Des.* **11** (1), 69-76. [39]

Michiyoshi, I, Serizawa, A and Kataoka, I (1974) Transport properties of air-water bubble flow in pipe. *5th Int. Ht. Transfer Conf. Presentation at Round Table Discussion on Momentum and Heat Transfer*, Sept 1974. [122,133]

Michiyoshi, I, Funakawa, H, Kuramoto, C et al (1977) Local properties of vertical mercury-argon two-phase flow in a circular tube under transverse magnetic field. *Int. J. Multiphase Flow.* **3** (5), 445-457. [122,136,154]

Mie, G (1908) Beitrage zur optik truben medien; speziell kolloidaler metallosungen. *Ann Physik.* **25**, 377-445. [154]

Miller, N and Mitchie, R E (1969) The development of a universal probe for measurement of local voidage in liquid-gas two-phase flow systems. *Eleventh Nat. ASME/AIChE Ht. Transfer Conf. Minneapolis*, 82-88. [125]

Miller, N and Mitchie, R E (1970) Measurement of local voidage in liquid-gas two-phase flow systems using a universal probe. *J. Brit. Nucl. Energy Soc.* **9**, 94-100. [125]

Milliot, B, Lazarus, J and Navarre, J Ph (1967) Void-fraction measurements in a two-phase flow. NaK/Argon, *Ispra-Italy*. EUR 3486f. [49,57,63]

Miropolsky, Z L (1955) The use of gamma rays. *Izv-Akad-Nank-SSR-Otdel-Tekh-Nauk*, **9**, 154-159. [49]

Miropolsky, Z L and Shneyerova, R I (1960) Heat engineering and hydrodynamics. 1. Measuring the volumetric content of steam - generating elements by means of gamma radiation. *USAEC* Report No AEC-tr-4206, 1. [49]

Miropolsky, Z L and Shneyerova, R I (1962) Application of X-rays, excited by β-sources, to studying hydrodynamics of two-phase media. *Inst. J. Heat Mass Transfer*, **5**, 723-728. [51,55]

Misch, F H (1963) Acoustical measurement of incipient boiling. *Inst. Engng. Res. Univ. California.* Report No NE-63-1. [76]

Mitzushina, T (1971) The electro-chemical method in transport phenomena. *Advances in heat transfer.* **7**, 87. [45]

Moeck, E O (1969) Measurement of liquid film flow and wall shear stress in two-phase flow. *11th Nat. ASME/AIChE Ht. Transfer Conf. Minneapolis*, 36-46. [161,171]

Moeck, E O, Garg, S C and Wikhammer, G A (1966) Swift dryout for a 19-rod, 3.25 in diameter bundle, cooled by steam-water fog at 515 psia. *Atom. Energy Canada.* Report AECL-2586. [66]

Moeck, E O and Stachiewicz, J W (1970) Liquid film behaviour in annular dispersed flow at critical heat flux. *4th Int. Ht. Transfer Conf. Paris, 1970.* [171]

Molen, S B and van der Galjee, F W B M (1977) Boundary layer and burnout phenomena in a subcooled two-phase flow. *European Two-Phase Flow Group Mtg, Grenoble, June 1977.* Paper No E1. [189]

Moore, F D and Mesler, R B (1961) The measurement of rapid surface temperature fluctuations during nucleate boiling of water. *AIChE Journal*, **7**, 620-624. [188]

Morgan, C D, Roy, D B and Hedrick, R A (1972) Analytical and experimental investigation of heat transfer during simulated cold-leg blow-down accidents. *Proc. of Crest Specialist Meeting on*

Emergency Cooling for Light Water Reactors. MRR 115, 2, Munich. [78]

Mori, Y, Hijikata, K and Kuriyama, I (1977) Experimental study of bubble motion in mercury with and without a magnetic field. *J. Heat Transfer,* **99** (3), 404-410. [122,133]

Morin, R (1965) Wall-temperature fluctuations during bubble generation in boiling. *Proc. of Symp. on Two Phase Flow, Exeter.* paper D3, 301-316. [188]

Morooka, S, Nishinaka, M and Kato, Y (1977) Overall mass transfer coefficient between the bubble phase and the emulsion phase in free and eight-stage fluidised beds. *Int. Chem. Eng.* **17** (2), 254-260. [44]

Morrison, G L (1974) Effects of fluid property variations on the response of hot-wire anemometers. *Journal of Physics E: Scientific Instruments,* **7**, 434-436. [128]

Morrow, T B and Kline, S J (1974) The performance of hot-wire and hot-film anemometers used in water, *In* "Flow - its measurement and control in science and industry", **1**, Part 2 (ISA), 555-562. [128]

Morton, J B and Clark, W H (1971) Measurement of two-point velocity correlations in a pipe flow using laser anemometers. *Journal of Physics E: Scientific Instruments,* **4**, 809-814. [135]

Moxon, D and Edwards, P A (1967) Dryout during flow and power transients. *AEEW-R553.* [79]

Mukhachev, G A and Susekov, O F (1972) Study of the speed of sound in gases. *Fluid Mechanics - Soviet Research.* **1** (3), 161-170. [71]

Mumme, I A and Lawther, K R (1973) The utilisation of fluid temperature fluctuations to assess the thermal performance of heat exchange equipment. *1st Australian Conference, Heat Mass Transfer. Monash U. May 1973.* Section 4.4, 33-38. [39]

Murdock, J W and Fiock, E F (1950) Measurement of temperature in high velocity steam. *ASME Trans.* **22**, 1155. (166]

Nabizadeh, H (1975) Experimental investigation to examine the void fraction in analogous water and Freon-113 flows. *Reactor Conf. Neurnberg,* 62-65. [203]

Nabizadeh, H (1977) Transfer laws between water and Freon-113 for mean void fraction, pressure drop and critical heat flux. *European Two-Phase Flow Group Mtg, Grenoble, June 1977.* Paper No E6. [205]

Nakaike, Y, Tadenuma, Y and Sato, T (1971) An optical study of interfacial turbulence in a liquid-liquid system. *Int. J. Heat Mass Transfer,* **14**, 1951-1961. [142]

Nariai, H (1977) Thermo-hydraulic behaviour of fluid in pressure vessel during blowdown. *Conf. on Heat and Fluid Flow in Water Reactor Safety, Manchester, Sept 1977.* 113-117. [72]

Nassos, G P (1963) Development of an electrical resistivity probe for void-fraction measurements in air-water flow. *Argonne Nat. Lab.* Report No ANL-6738. [122]

Nassos, G P (1965) Propagation of density disturbances in air-water flow. *ANL* 7053. [122]

Nassos, G P and Bankoff, S G (1967) Local resistivity probe for study of point properties of gas-liquid flows. *Canad. Jnl. Chem. Eng.* **45**, 271-274. [122]

National Engineering Laboratory (1966) Metering of two-phase mixtures. *Nat. Engng. Lab. UK.* Report No 217.

Navon, U and Fenn, J B (1971) Interfacial mass and heat transfer during evaporation. 1. An experimental technique and some results with a clean water surface. *AIChE Journal.* **17** (1), 131-140. [141]

Neal, L G and Bankoff, S G (1963) A high resolution resistivity probe for determination of local void properties in gas-liquid flow. *AIChE Journal.* **9**, 490-494. [122,133]

Neal, L G and Bankoff, S G (1965) Local parameters in cocurrent mercury-nigrogen flow (Pts 1 and 2) *Symp. on Two-Phase Flow, Exeter, 1965.* Paper C6. [122,133]

Neal, S B H (1975) The development of the thin-film naphthalene mass-transfer analogue technique for direct measurement of heat-transfer coefficients. *Int. J. Heat Mass Transfer,* **18** (4), 559-567. [44]

Neal, S B H C, Northover, E W and Hitchcock, J A (1970) The development of a technique for applying naphthalene to surfaces for mass transfer analogue investigations. *J. Phys. E. Sci. Instrum.* **3**, 636-638. [44]

Nedderman, R M (1961) The use of stereoscopic photography for the measurement of velocities in liquids. *Chem. Engng. Sci.* **16**, 113-119. [131]

Neomi, S and Rose, J W (1977) Heat-transfer measurements during dropwise condensation of mercury. *Int. J. Heat Mass Transfer.* **20** (8), 877-881. [35]

Nerem, R M and Stickford, G H (1964) A thin-film radiative heat-transfer cage. *AIAAJ.* **2**, 1647-1651. [38]

Nevstruyeva Ye I, Romanovsbiy, I M and Styrikovich, M (1973) Analysis of certain phenomena accompanying the boiling process. *Heat Transfer Sov. Res.* **5** (4), 91-96. [77]

Newby, C A (1971) Heat transfer to falling films. *M. Sc. Thesis, U. Manchester, 1971.* [37]

Nibler, J W, McDonald, J R and Harvey, A B (1976) CARS measurement of vibrational temperatures in electric discharges. *Optics Communications,* **18**, 71. [145,168]

Nicol, A A, Bryce, A and Ahmed, A S A (1978) Condensation of a horizontally flow-ng vapour on a horizontal cylinder normal to the vapour stream. *Proceedings of the 6th International Heat Transfer Conference, Toronto, August 1978.* **2**. [35]

Nicolitas, A S and Murgatroyd, W (1968) Precise measurement of slug speeds in air-water flows. *Chem. Eng. Sci.* **23** (8), 934-936. [139]

Nigmatulin, B I, Malyshenko, V I and Shugaev, Y Z (1977) Investigation of liquid distribution between the core and the film in annular dispersed flow of steam/water mixture. *Therm. Eng.* **23** (5), 66-68. [171]

Nijaguna, B T and Abdelmessih, A H (1974) Precoalescence drop growth-rates in dropwise condensation. *5th Internat. Ht. Transfer. Conf. Tokyo, Sept 1974.* Paper Cs2.3, 264-268. [193]

Nishikawa, N K, Nishi, N, Sekoguchi, K and Nakasatomi, M (1967) Liquid film flow phenomena in upwards two-phase annular flow. *Jap. Soc. Mdch. Eng. Semi. Inter. Symp.* Paper 260, 65-74. [177]

Nishikawa, K, Sekoguchi, K and Fukano, T (1969) On the pulsation phenomena in gas-liquid two-phase flow. *Bull. JSME,* **12** (54), 1410-

1416. [185]

Nishikawa, K, Sekoguchi, K, Nakasatomi, M and Kaneuzi, A (1968) Two-phase annular flow in a smooth tube and grooved tubes. *Int. Symp. Res. On cocurrent Gas-Liquid Flow. Univ. of Waterloo, Canada, Sept 1968*. [177]

Nishio, S and Hirata, M (1978) Direct contact phenomenon between a liquid droplet and high temperature solid surface. *Proceedings of the 6th International Heat Transfer Conference, Toronto, August 1978*. **1**, 245. [83]

Nodack, R (1970) Local heat transfer at horizontal tubes in fluidised beds. *Chem. Ing. Tech.* **42** (6), 371-376. [38]

Nones, F P and Felton, G L (1973) Accurate flow measurement with target flow transmitters. *Advances in Instrumentation*. **28** (4).

Northover, E W and Hitchcock, J A (1964) *Thermodynamics and Fluid Mech. Convention. Inst. Mech. Engrs. Cambridge*. Paper 2, 3-8. [39]

Nychas, S G, Hershey, H C and Brodkey, R S (1973) A visual study of turbulent shear flow. *J. Fluid Mech.* **61**, Part 3, 513-540. [131]

Nyer, L J et al (1967) Propagation of shock waves through two-phase, two-component media. *ANS Trans.* **10** (2), 660. [71,72]

Nylund, O, Becker, K M, Eklund, R and Gelius, O, et al (1968). Hydrodynamic and heat transfer measurements on a full scale simulated 36-rod marviken fuel element with uniform heat flux distribution. *Europ. Two-Phase Flow Mtg. Oslo, June 1978. AB Atomenergi, Sweden*. Paper D2.1, Report No FRIGG-2. [121]

Nylund, O, Gelius, O and Rouhani, Z (1968) Full-scale loop studies of the hydrodynamic behaviour of BHWR fuel. *European Two-Phase Group Meeting, Oslo, June 1968*. [121]

Ochial, M A, Furukawa, K, Kuroyanagi, T and Kobayashi, K (1972) Void fractions and heat transfer coefficients for sodium-argon in two-phase flow in a vertical channel. *Heat Transfer Jap. Res.* **1**, Pt 2, 65-73. [49]

Ohba, K (1974) Simultaneous measurement of local flow velocity and void fraction in bubbly flows using a gas laser. *5th Int. Heat Transfer Conf. Presentation made at the Round Table Discussion on Momentum and Heat Transfer Mechanism in Two-Phase Flow, Sept 1974*. [138]

Ohki, Y and Inoue, H (1970) Longitudinal mixing of the liquid phase in bubble columns. *Chem. Engng. Sci.* **25** (1), 1-16. [149]

Ohta, Y, Shimoyama, K and Ohigashi, S (1975) Vaporisation and combustion of single liquid fuel droplets in a turbulent environment. *Bull. JSME*. **18** (115), 47-57. [43]

Oka, S, Kostic, Z and Pislar, V (1970) Transient method for measurement of local heat transfer coefficient on extended heated surfaces. *Bull. Boris Kidric Inst. Nucl. Sci.* **2** Nucl. Eng'g (2) 461. [39]

Oker, E and Merte, H (1978) A study of transient effects leading up to inseption of nucleate boiling. *Proceedings of the 6th International Heat Transfer Conference, Toronto, August 1978*. **1**, 139. [192]

Oki, K, Walawendor, W P and Fan, L T (1977) The measurement of local velocity of solid particles. *Powder Technol*. **18** (2), 171-178. [139]

Oldengarm, J, Krieken, A H van der Raterink, H J (1973) Laser-Doppler velocimeter with optical frequency shifting. *Optics Laser Technol.* **5** (6), 249-252. [137]

Oldengarm, J, Krieken, A H van der Raterink, H J (1975) Velocity profile measurements in a liquid film flow, using the laser Doppler technique. *J. Phys. E. Sci. Instrum.* **8** (3), 203-205. [138]

Oliver, D R and Young Hoon, A (1968) Two-phase non-newtonian flow. Part 1: Pressure drop and hold-up. *Trans. Instn. Chem. Engrs.* **46**, T106-T115. [31,61]

Olsen, H O (1967) Theoretical and experimental investigation of impedance void meters. *Kjeller, Norway,* Report 118. [58,59]

Olszowski, S T, Coulthard, J and Sayles, R S (1976) Measurement of dispersed two-phase gas-liquid flow by cross correlation of modulated ultrasonic signals. *Int. J. Multiphase Flow,* **2** (5-6), 537-548. [141]

Orbeck, I (1962) Impedance void meter. Institutt for Atomenergie, *Kjeller Research Establishment, Norway.* KR32. [57]

Osborne, M F M and Holland, F H (1947) The acoustical concomitants of cavitation and boiling produced by a hot wire. *J. Acoust. Soc. Amer.* Part 1, **19**, 13-20. [76]

Padmanabhan, M (1976) Wave propagation through flowing gas-liquid mixtures in long pipelines. A thesis presented to the faculty of the division of graduate studies. *Ph.D Thesis, Geogia Inst. Tech.* [71]

Paleev, I I and Filippovich, B S (1966) Phenomena of liquid transfer in two-phase dispersed annular flow. *Int. Jnl. Heat Mass Transfer.* **9**, 1089-1093. [169]

Palm, J W, Kirkpatrick, JW and Anderson, W H (1968) Determination of steam quality using an orifice-meter. *Jnl. Petrol. Techn.* **20**, 587-590. [96]

Papadatos, K and Svrcek, W Y (1974) Low-range flowmeter for pulverised solids. *Instr. Techn.* 38-40 [130]

Park, J Y and Blair, L M (1975) The effect of coalescence on dropsize distribution in an agitated liquid-liquid dispersion. *Chem. Eng. Sci.* **30**, 1057-1064. [189]

Park, R W and Crosby, E J (1965) A device for producing controlled collisions between pairs of drops. *Chem. Engng. Sci.* **20**, 39-45. [192]

Parker, G J (1968) Some factors governing the design of probes for sampling in particle- and drop-laden streams. *Atmos. Environment,* **2**, 477-490.

Parker, G J and Ryley, D J (1969-70) Equipment and techniques for studying the deposition of sub-micron particles on turbine blades. *Proc. Instn. Mech. Engrs.* **184**, Part 3C, 43-51. [150]

Parker, R (1965) Method for determining the response of temperature sensors to a rapid temperature rise. *Review. Sci. Instr.* **36**. [

Parsons, C B and Brundrett, E (1972) A combined photographic filmthickness and temperature technique for observing dryout. *Int. Symp. Two-Phase Systems, Haifa, Israel, Aug-Sept 1971.* Session 3, Paper 10, *Progress in Heat and Mass Transfer.* **6**, 365-384. [67]

Pastorius, W J and Pryor, T R (1974) Diffractographic techniques for vibration measurement. *J. Engng. Industr.* **96** (2), 553-556. [86]

Patent No 1,230,159. (Filed 11 Nov 1968). [63]
Pekrul, P J (1970) Transit time flow meter. *Quarterly Tech. Progress Report, LMFBR Task 25*. AI-AEC-12981 (Instruments). [132]
Perkins, A S and Westwater, J W (1956) Measurements of bubbles formed in boiling methanol. *AIChE Journal*, 2, 471-476. [152]
Perkins, H C, Yusuf, M and Leppert, G (1961) A void measurement technique for local boiling. *Nucl. Sci. and Engng*. 11, 304-311. [54]
Permyakov, V V and Podsushniy, A M (1971) Liquid-film annulus depth and stability in cocurrent gas-liquid flow. *Heat Transfer-Soviet Research*, 3 (2). [114]
Persson, S L (1964) Method for determination of velocity distribution in a thin liquid film. *AIAA Journal*, 2, Part 2, 372-373. [131]
Petrick, M and Swanson, B S (1958) Radiation attenuation method of measuring density of a two-phase fluid. *Rev. Scient. Instrum*. 29, (12), 1079-1085. [
Petrick, M and Swanson, B S (1959) Expansion and contraction of an air-water mixture in vertical flow. *AICHE Jnl*. 5 (4), 440. [49,5), 52,53]
Petrunik, K H and St Pierre, C C (1970) Turbulent Mixing rates for air-water two-phase flows in adjacent rectangular channels. *Canad. Jnl. Chem. Engng*. 48, 123-125. [149]
Pettigrew, M J (1977) Flow-deduced vibration of nuclear power station components. *AECL Report No* AECL 5852. [86]
Pettigrew, M J and Gorman, D J (1977) Experimental studies on flow-induced vibration to support steam generator design. Part III: Vibration of small tube bundles in liquid and two-phase cross-flow. *AECL* Report No 5804. [86]
Pettigrew, M J and Patdoussis, M P (1975) Dynamics and stability of flexible cylinders subjected to liquid and two-phase axial flow in confined annuli. *Chalk River Nucl. Labs*. Report No. AECL-5502. [86]
Piggott, B D G and Porthouse, D T C (1973) A correlation of rewetting data. *Private communication, CEGB, Berkeley Laboratory*. [84]
Pike, J G (1965) Influence of a falling thin liquid film upon a co-currently flowing gas-stream in a vertical duct. *Canad. Jnl. Chem. Engng*. 43, 267-270. [128]
Pike, R W, Jackson, D A, Bourke, P J and Page, D I (1968) Measurement of turbulent velocities from the Doppler shift in scattered laser light, *Journal of Physics E: Scientific Instruments*, 1, Series 2, 727-730. [135]
Pike, R W, Wilkins, B and Ward, H C (1965) Measurement of the void fraction in two-phase flow by X-ray attenuation. *AIChE Journal*, 11, 794-800. [51]
Pimsner, V and Toma, P (1977) The wavy aspect of a horizontal co-current air-water film flow and the transport phenomena. *Int. J. Multiphase Flow*. 3, 273. [160]
Pinczewski, W V and Fell, C J D (1972) The transition from froth-to-spray regime on commercially loaded sieve-trays. *Trans. Inst. Chem. Engrs*. 50 (2), 102-108. [107]
Piper, T C (1971) Final report on the semi-scale gamma attenuation two-phase water density measurement. *Microfiche* No IN-1487. [49]

Piper, T C (1974a) Dynamic gamma attenuation density measurements. *Aerojet Nuclear Company*, ANCR-1160. [51]

Piper, T C (1974b) Small turbines for fuel pin channel flow distribution measurements. Nuclear Technology Division Annual Progress Report for Period Ending June 30, 1974. *ANCR*-1177, 284-287. [98]

Piplies, L (1972) Velocity of solids in gas/solid flow in vertical pipes (in German, English abstract). *Chemie Ing. Techn.* **44** (8), 394-399. [139]

Pitts, D R, Hewitt, H C and McCollough, B R (1977) Heat transfer controlled collapse of a cylindrical vapour bubble in a vertical isothermal tube. *J. Heat Transfer.* **99** (3). [77]

Pletcher, R H and McManus, H N (1969) Heat transfer and pressure drop in horizontal annular two-phase, two-component flow, (and letter). *Int. Jnl. Heat Mass Transfer*, **11**, 1087-1104 (1968) and **12**, 663-669 (1969). [37]

Pogson, J T, Roberts, J H and Waibler, P J (1970) An investigation of the liquid distribution in annular-mist flow. *Jnl. Heat Transfer*, **92**, Series C, 651-658. [152]

Ponter, A B, Davies, G A, Ross, T K and Thornley, P G (1967) The influence of mass transfer on liquid film break-down. *Int. J. Heat and Mass Transfer.* **10**, 349-359. [84]

Ponter, A B and Haigh, C P (1969) Sound emission and heat transfer in low pressure pool boiling. *Int. J. Heat Mass Transfer*, **12**, 413-428. [76]

Poole, D H (1970) An ultrasonic technique for measuring transient movements of a phase boundary in a liquid, *Journal of Physics E: Scientific Instruments*, **3**, 726-823. [78]

Popovich, A T and Hummel, R L (1967) A new method for non-disturbing turbulent flow measurements very close to a wall. *Chem. Eng. Sci.* **22**, 21-25. [133]

Popper, J, Abauf, N and Hetsroni, G (1974) Velocity measurements in a two-phase turbulent jet. *Int. J. Multiphase Flow*, **1** (5), 715-726. [138]

Portalski, S and Clegg, A J (1972) An experimental study of wave inception on falling liquid films. *Chem. Engng Sci.* **27**, 1257-1265. [178]

Possa, G, Valli, G and van Erp, J B (1965) Measurement of total steam volume in a heated channel at 70 Kg/cm^2. *Symp. Two-Phase Flow, Exeter*, Paper E4. [61]

Possa, G and van Erp, J B (1965) Measurement and analysis of two-phase flow dynamics. *Symp. Two-Phase Flow, Exeter*. Paper H6. [92]

Prasad, C, Chen, C S and Beard, J T (1971) Analytical and experimental studies of heat and mass transfer from an air-water interface. *Int. Symp. Two-Phase Systems, Israel*, Paper 3-14, Session 3. [141]

Premoli, A and Hancox, W T (1976) An experiemntal investigation of subcooled blowdown with heat addition. *CSNI Paper presented Specialist Mtg. Transient Two-Phase Flow. Toronto, Aug 1976*. [72]

Preston, J H (1954) The determination of turbulent skin friction by means of Pitot tubes. *J. Roy. Aero. Soc.* **58**, 108-121. [165]

Price, I and Bramall, J W (1971) A diametrical effect on vapour column formations in film boiling in carbon dioxide near the critical state. *Int. J. Heat Mass Trasffer.* **14**, 1750-1751. [193]

Prisco, M R, Henry, R E, Hutcherson, M N et al (1977) Non-equilibrium critical discharge on saturated and subcooled liquid Freon-11. *Nucl. Sci. Eng.* **63** (4), 365-375. [70]

Puzyrewski, R and Jasinski, R (1965) Measurement of the thickness of thin water-film by resistance method. *Proc. Inst. Turbomachinery (Gdansk, Poland)*, **26**, 73-83. [112]

Pye, J W (1971) Droplet size distribution in sprays using a pulse-counting technique. *J. Inst. Fuel.* 253-256. [153]

Quandt, E R (1962) Measurement of some basic parameters in two-phase annular flow. *AIChE Symp. Two-Phase Flow, Chicago.* Report No WAPD-T-1502. [146]

Quinn, E P and Swann, C L (1964) Visual observations of fluid behaviour in high-pressure transition boiling flows. *General Electric Report.* GEAP-4636. [189]

Raasch, J and Umhauer, H (1977) Fundamental consideration in the measurement of particle size and particle velocity of dispersed phases in flow systems, *Chem. Ing. Tech.* **49** (12), 93-1941. [138, 185].

Raisson, C (1965) Flow regime studies up to critical heat-flux conditions at 80 kg/cm^2. *CEN Grenoble*, Report No TT62. [106,109]

Rakapoulos, C D, El-Shirbini, A A and Murgatroyd, W (1978) An experimental investigation into the dynamics of two-phase flow in vapour generators. *Proceedings of the 6th International Heat Transfer Conference, Toronto, August 1978.* **1**, 357. [92]

Ralph, J C, Sanderson, S and Ward, J A (1977) Experimental studies of post-dryout heat transfer to low quality steam-water mixtures at low pressures. *European Two-Phase Flow Group Mtg, Grenoble. June 1977.* Paper No C3. [39,40]

Ralph, J C, Tomlinson, L, Silver, P J B and Lilley, W (1978) Studies of waterside corrosion and heat transfer on LMFBR evaporator tubes. *Proceedings of the 6th International Heat Transfer Conf. Toronto, August 1978.* **4**, 379. [39]

Ram, K S and McKnight, R D (1970) Noise analysis techniques as applied to nucleate pool boiling. *4th Int. Heat Transfer Conf. Paris, 1970.* **5**, Paper B1.7. [76]

Randall, R L (1969) Transit time flowmeter employing noise analysis techniques, Part I, Water loop tests. *Atomics International*, AI-AEC-12802. [132]

Randall, R L (1970) Transit time flowmeter employing noise analysis techniques, Part II, 2-In. Sodium-loop tests. *Atomics International* AI-AEC-12941. [132]

Rao, C S and Dukler, A E (1971) The isokinetic-momentum probe. A new technique for measurement of local voids and velocities in flow of dispersions. *Industr. Eng. Chem. Fundam.* **10** (3), 520-526. [129]

Rasmussen, C G and Jensen, J D (1971) Laser-Doppler anemometry. A signal processing problem. *Int. Jymp. Two-Phase Systems, Haifa, Israel, Aug-Sept 1971.* Session 7, Paper 1. [138]

Rasmussen, L and Malnes, D (1974) Simulation of blow-down from pressure vessels based on non-thermal equilibrium effects. Comparison with experimental data. *Pres. at European Two-Phase Flow Group Meeting, Harwell, 1974.* [72]

Rea, J and Chojnowski, B (1974) Voidage in a high-pressure steam-generating tube. *Inst. Chem. Engrs. Symp. No 38*, 2, Paper No 14. [62]

Reddy, K V S, van Wijk, M C and Pei, D C T (1969) Stereophotogrammetry in particle-flow investigation. *Canad. Jnl. Chem. Engng.* 47, 85-88. [132]

Regnier, P (1973) Application of coherent anti-Stokes Raman scattering to the measurement of gas concentrations and to flow visualisation. *European Space Agency.* Report ESA-TT-200. [145]

Regnier, P, Moya, F and Taran, J P E (1974) Gas concentration measurement by coherent Raman anti-Stokes scattering. *AIAA Journal*, 12, 826. [145]

Reimann, J and John, H. (1978) Measurements of the phase distribution in horizontal air-water and steam-water flow. *Paper presented at CNSI Meeting on Transient Two-Phase Flow, Paris, June 1978.* [107, 109,122]

Reisbig, R L (1974) Macroscopic growth mechanisms in dropwise condensation. *5th Internat. Ht. Transfer Conf. Tokyo, Sept 1974.* 255-258. [193]

Reith, T (1970) Interfacial area and scaling-up of gas-liquid contractors. *Brit. Chem. Engng.* 15, 1559-1563. [154]

Reocreux, M and Flamand, J C (1972) The use of electrical probes in the study of two-phase high-speed flows. *CEN Grenoble,* Report No TT 111. [133]

Resch, F (1968) Etudes sur le fil chaud et la film chaud dans l'eau. *CEA (Saclay)* CEA-R-3510. [128]

Rhodes, R G and Scott, D S (1975) Studies of gas-liquid (non-Newtonian) slug flow: void fraction meter, void fraction and slug characteristics. *Chem. Engng.* 10 (1), 57-64. [

Richardson, B L (1958) Some problems in horizontal two-phase, two-component flow. *Argonne Nat. Lab.* Report No ANL-5949. [49,53]

Riddle, J L et al (1973) Platinum resistance thermometry. *NBS* Monograph 126. [167]

Riebold, W, de Cooman, F and Friz, G (1970) The application of a correlation method in two-phase flow investigations. *European Two-Phase Flow Group Meeting, Milan, 1970.* Paper C5. [139,140]

Riedle, K and Winkler, F (1972) EEC - reflooding experiments with a 340-rod bundle. *Crest Specialist Meeting on Emergency Core Cooling for Light Water Reactors, Munich.* MRR 115, 1. [82]

Riegel, B (1971) The effect of a cold wall and the influence of a diaphragm in an annular cold channel with Freon-12. *CEN Grenoble,* Report No TT 105. [203]

Rinne, G (1975) Relationship between steady-state dryout conditions for water and Freon-12. *Reactor Conf. Neurnberg.* 58-61. [205]

Rizzo, J E (1975) A laser-Doppler interferometer. *Journal of Physics E: Scientific Instruments.* 8, 47-52. [137]

Robertson, A D and Sheridan, A T (1970) Water modelling of liquid steel at Swinden Laboratories. *Jnl. Iron Steel Inst.* 625-632. [196]

Robinson, G E, Schmidt, F W, Block, H R and Green, G (1974) An experimental determination of isolated bubble acoustics in a nucleate boiling system. *5th Internat. Ht. Transfer Conf. Sci. Council. Japan, Sept 1974.* 4. [76]

Roemer, R B (1970) Surface temperature fluctuations during steady state boiling. *Int. Jnl. Heat Mass Transfer.* **13**, 985-996. [186]

Rogers, T F and Mesler, R B (1964) An experimental study of surface cooling by bubbles during nucleate boiling of water. *AIChE J.* **10** (5), 656-660. [188]

Rooney, D H (1972) Steam-water flow through orifices. *Nat. Engng. Lab. UK.* Report No 549. [96]

Rose, S and Griffith, P (1964) Bubble and froth in vertical pipes. Two-phase gas-liquid flow, *Summer Program, Mass. Inst. Techn. 1964.* [161]

Rosehart, R G D and Jagota, A E (1968) A method for the determination of individual mass transfer coefficients and interfacial areas for the liquid film and droplets in cocurrent gas-liquid annular flow. *Int. Symp. Res. in Cocurrent Gas-Liquid Flow, U. Waterloo, 1968.* **2**. [146]

Rosehart, R G D, Rhodes, R G and Scott, D S (1975) Studies of gas-liquid (non-newtonian) slug flug: void fraction meter, void fraction and slug characteristics. *Chem. Engng.* **10** (1), 57-64. [57,182]

Ross, P A and El-Wakil, M M (1960) A two-wavelength interferometric technique for the study of vaporisation and combustion of fuels. *Progress in Astronautics and Rocketry.* **2** (2), 265-298. [142]

Rothe, P H (1977) PWR steam generator waterhammer technology. *Paper presented at the NRC 5th Water Reactor Safety Research Information Meeting, Gaithersburg, Maryland, USA, November 1977.* [78]

Roughton, J E and Crosse, P A E (1974) A 'thermal transient' instrument for the measurement of flow in pipes by external means. *Journal of Physics E: Scientific Instruments.* **7**, 621-626. [132]

Rouhani, S Z (1962) Void measurement by the (γ,n) reaction. *Nucl. Sci. and Engng.* **14**, 414-419. [55]

Rouhani, S Z (1964) Application of the turbine type flowmeters in the measurement of steam quality and void, *Symposium on In-Core Instrumentation, Oslo, June 1964.* [98]

Rouhani, S Z (1965) Void measurements in the region of sub-cooled and low-quality boiling. *Symp. Two Phase Flow, Exeter,* Paper E5. Also AE-238. [55]

Rouhani, S Z (1974) Measuring techniques. *Von Karman Inst. Fluid Dynamics. Lecture Series 71. Two-Phase Flows with Application to Nuclear Reactor Design Problems. Dec 1974.* [98]

Rouhani, S Z and Becker, K M (1963) Measurements of void-fractions for flow of boiling heavy water in a vertical round duct. *Aktiebolaget Atomenergie* Report No AE-106. 2nd Rev. Ed. [55]

Roukens de Lange, A (1975) Dispersion in annular climbing film flow. *Ph.D Thesis, U. Witwatersrand, Johannesburg.* [148]

Roumy, R (1969) Structure of two-phase air-water flows: study of average void fraction and flow patterns. *CEA*-R-3892. [122]

Roumy, R (1970) Dry-out with boiling Freon-12 in straight tubes of different slops. *European Two-Phase Flow Group Meeting, Milan, June 1970.* Paper D13. [61,203]

Rousseau, J C, Czerny, J and Riegel, B (1976) Void fraction measurements during blowdown by neutron absorption of scattering methods. *European Two-Phase Flow Meeting, Erlangen, June 1976.* Paper No B5. [55]

Rousseau, J C and Riegel, B (1978) Super-CANON experiments. *Paper presented at CNSI Specialist Meeting on Transient Two-Phase Flow, Paris, June 1978.* [55,72]

Rowe, P N and Evans, T J (1974) Dispersion of tracer gas supplied at the distributor of freely bubbling fluidised beds. *Chem. Engng. Sci.* **29**, 2235-2246. [149]

Rubin, I R and Roizen, L I (1974) Mechanism of pool boiling heat transfer of water on non-isothermal surfaces. *5th Internat. Heat Transfer Conf. Sci. Council Japan, Sept 1974.* **4**. [122]

Rudenko, M I (1971) Pickup for local heat fluxes. *High Temp.* **9** (4), 776. [38]

Rudiger, B (1972) Experimental blow-down studies of a reactor simulator vessel containing simplified internals. *Proc. of Crest Specialist Meeting on Emergency Core Cooling for Light Water Reactors, Munich.* MRR 115, **1**. [72,73]

Russel, T W F and Lamb, D E (1965) Flow mechanism of two-phase annular flow. *Canad. Jnl. Chem. Engng.* **43**, 237-245. [149]

Rutz, J [1971] Techniques for measuring wetness of steam. *Kernenergie,* **14** (6), 182-187. [96]

Ryan, J T and Vermeulen, L R (1971) A slug-flow transducer. *Review Sci. Instr.* **39**, 1756-1757. [107]

Ryley, D J (1966) Wet steam property measurements: current problems: a review. *Int. Jnl. Mech. Sci.* **8**, 581-589. [96]

Ryley, D J and Fallon, J B (1966) Size sampling of steam-borne water droplets. *Thermodynamics Fluid Mech. Conv. Liverpool, April 1966.* Paper 25. [153]

Ryley, D J and Holmes, M J (1973) Sampling of high-quality wet steam from steam mains operating at 11.4 bar pressure. *Proc. Instn. Mech. Engrs.* **187**, 381-393. [96]

Ryley, D J and Kirkman, G A (1967-68) The concurrent measurement of momentum and stagnation enthalpy in a high-quality wet steam flow. *Proc. Instn. Mech. Engrs.* **182**, Pt 3H, 250-257. [129]

Ryley, D J, Ralph, W J and Tubman, K A (1970) The collision behaviour of water drops within a low-pressure steam atmosphere. *Int. Jnl. Mech. Sci.* **12**, 589-596. [192]

Ryzhkov, S V and Khmara, O M (1978) Laser anemometry investigation of two-phase wall layers in triangular ducts. *Proceedings of the 6th International Heat Transfer Conference, Toronto, August 1978.* **1**, 481. [185]

Samoilovich, G S and Yablokov, L D (1970) Measurement of periodically-fluctuating flows in turbo-machines by ordinary pitot tubes. *Thermal Engng.* **17** (9), 105-110. [128]

Sandervag, C (1971) Thermal non-equilibrium and bubble size distributions in an upward steam water flow. *Inst. Atomenergi, Kjeller, Norway,* Report No KR-144. [122]

Sato, Y (1974) An experimental investigation of air-bubble motion in water-streams in a vertical duct. *5th Int. Heat Transfer Conf. Presentation at Round Table Discussion on Momentum and Heat Transfer, Sept 1974.* [124,189]

Saxe, R F (1965-66) Detection of boiling in water-moderated nuclear reactors. *Nuclear Safety,* **7** (4), 452-460. [76]

Saxe, R F (1969) Considerations regarding the acoustical detection of boiling in the presence of cavitation. *North Carolina State*

Univ. Dept of Nucl. Engng. Report No 27607. [76]
Saxe, R F and Lau, L W (1968) Cavitation noise in nuclear reactors. *Nucl. Engng. and Design.* **8**, 229-240. [76]
Saxe, R F, Sides, W H and Foster, R G (1971) The detection of boiling in nuclear reactors. *Jnl. of Nucl. Engng.* 139-153. [76]
Schenk, K (1967) Development of an in-pile void gauge. Inst. Atomenergi, *OECD Halden Reactor Project.* Report No HPR 75. [57]
Schlapbach, M E (1977) Determination of condensing heat transfer coefficients and two-phase flow regimes for dichlorotetrafluoroethane refrigerant (R-114). *Union Carbide* Report No KY-674. [106]
Schmidt, E W, Boiarski, A A, Gieseke, J A et al (1976) Applicability of laser interferometry technique for drop size determination. *Paper presented Symp. American Chem. Soc. San Francisco, Aug-Sept 1976.* [159]
Schmidt, F W, Robinsin, G E and Skapura, R J (1970) Experimental study of noise generation in a nucleate boiling system. *Heat Transfer*, **5**, Paper B 1.8. [76]
Schmidt, H (1977) Fast co-axial thermocouples. *Kerntechnische Gesellschaft im Deutschen Atomforum ev*, Proceedings of a Specialist Conference, Feb-March 1977, Hannover; "Experimental techniques in the field of thermodynamics and fluid dynamics. Part II; thermometry techniques and simulation of thermodynamics processes". 245. Interatom, Berg-Gladbach, West Germany. [77]
Schneiter, G R (1966) Mass transfer in annular two-phase flow. *Univ. Purdue*, Report No F-66-2. [42]
Schraub, F A (1966) Isokinetic sampling probe technique applied to two-component, two-phase flow. *General Electric AEC* R & D Report GEAP-5739. [61,114,165]
Schrock, V E (1969) Radiation attenuation techniques in two-phase flow measurement. *Eleventh Nat. ASME/AIChE Ht. Transfer Conf. Minneapolis, Aug 1969.* 24-35. [49]
Schrock, V E, Johnson, H A, Gopalakrishnan, A, Lavezzo, K E and Cho, S M (1966) Reactor heat transients, final report. Transient boiling phenomena, *SAN* 1013, **1** and **2**. [77]
Schrock, V E and Selph, F B (1963) Reactor heat transients research. An X-ray densitometer for transient steam-void measurement. *U. California. Inst. Engng.* Research Report No SAN-1005, TID-4500. [51,61,78]
Schulte, E H and Kohl, R F (1970) A transducer for measuring high heat-transfer rates. *Rev. Sci. Instrum.* **41** (12), 1732-1740. [38]
Schwanbom, E A, Braun, D and Hamann, E (1971) A double-ray technique for the investigation of liquid boundary layers. *Int. Jnl. Heat Mass Transfer.* **14**, 996-998. [182,183]
Scruton, B and Chojnowski, B (1978) Two-phase heat transfer in serpentine geometry boiler tubes at high pressure. *Proceedings of the 6th International Heat Transfer Conference, Toronto, Aug 1978.* **2**, 55. [34]
Sekoguchi, K (1968) The influence of mixers, bends and exit sections on horizontal two-phase flow. *Internat. Symp. Res. Cocurrent Gas-Liquid Flow. Univ. of Waterloo, Sept 1968.* [122,133]
Sekoguchi, K, Fukui, H, Matsuoka, T et al (1975) Investigation into the statistical characteristics of bubbles in two-phase flow. *Bull. JSME*, **18** (118). 391-396. [122]

Sekoguchi, K, Gukui, H, Tsutsui, M et al (1975) Investigation into the statistical characteristics of bubbles in two-phase flow (2nd report, application and establishment of electric resistivity probe method). *Bull. JSME.* **18** (118), 397-404. [133,153]

Sekoguchi, K, Nishikawa, K and Nakasatomi, M (1974) Flow-boiling in sub-cooled and low-quality regions-heat transfer and local void fraction. *5th Internat. Ht. Transfer Conf. Sci. Council, Japan, Sept 1974.* **4**. [122]

Semenov, N I (1959) Pressure pulsations during the flow of gas-liquid mixtures in pipes. Heat Power Engng. Pt 1, *AEC*-TR-4496, 58-65. [185]

Semenov, N I and Kosterin, S I (1964) Results of studying the speed of sound in moving gas-liquid systems. *Thermal Energy.* **11** (6), 59-64. [71]

Semeria, R and Flamand, J C (1967) Use of a micro-thermocouple for the study of local boiling of water under conditions of free convection. *CEN Grenoble.* Report No TT81. [186]

Semeria, R and Flamand, J C (1970) Analysis of subcooled boiling by micro-thermocouples. *European Two-Phase Flow Group Meeting, Milan, June 1970.* Paper C3. [186]

Serizawa, A (1974) Fluid-dynamic characteristics of two-phase flow. Inst. of Atomic Energy. *Ph.D Thesis, Kyoto Univ.* [26,61,133,134]

Serizawa, A, Kataok, A I and Michiyoshi, I (1975) Turbulence structure of air water bubbly flow 1. Measuring techniques. *Int. J. Multiphase Flow,* **12** (3), 221-233. [122,133]

Seynhaeve, J M (1977) Critical flow through orifices. *European Two-Phase Flow Group Mtg. Grenoble. June 1977.* Paper No B9. [70]

Sha, W T and Bonilla, C F (1965) Out-of-pile steam-fraction determination by neutron-beam attenuation. *Nuclear Applications.* **1**, 69-75. [54]

Shen, C N and Mihalek, E W (1977) Applications of a scattering spectrometer system for marine aerosol measurement. *J. Eng. Power,* **99** (4), 533-536.

Sheppard, J D, Leavell, W H, Shahrokhi, F and Wynn, M C (1977) Progress report on advanced two-phase instrumentation. *US Nuclear Regulatory Commission, Proceedings of Meeting of Review Group on Two-Phase Flow Instrumentation, January 1977.* NUREG-0375, (Paper No I.1). [96]

Sheppard, J J and Bradfield, W S (1971) Stagnation - point, free - convection film boiling on a hemisphere. *Internat. Symp. on Two-Phase Systems. Haifa, Israel, Aug-Sept 1971.* Paper 2-12. [177]

Sher, N C, Kangas, G and Neusen, K F (1962) On the phenomenon of boiling flow in a parallel rod array. *AIChE Mtg. Chicago, Dec 1962.* [49]

Shertok, J T (1976) Velocity profiles in core-annular flow using a laser Doppler velocimeter. *Ph.D Thesis, Princeton U.* [138]

Shih, T D (1974) Measurement of transient critical heat flux by fluid modelling. *Ph.D Thesis, Iowa State U.* [79,80]

Shilimkan, R V and Stepanek, J B (1977) Effect of tube size on liquid side mass transfer in co-current gas-liquid upward flow. *Chem. Eng. Sci.* **32** (11), 1397-1400. [41,42]

Shimamune, H, Shiba, M and Adachi, H (1972) Current status of ROSA program. *Proc. of Crest Specialist Meeting on Emergency Cooling for Light Water Reactors, Munich, 1972.* MRR 115, 2. [72]

Shimamune, H, Toda, S and Ookubo, H (1969) The effect of pressure trasients on thermal behaviours of heated metal foil during blowdown. *Crest Specialist Meeting, Depressurisation Effects in Water-Cooled Reactors, Frankfurt, 1969.* [79]

Shin, Y S and Wamsbganss, M W (1977) Flow-induced vibration in LMFBR steam generators. A state-of-the-art review. *Nucl. Eng. Design.* **40** (2), 235-284. [86]

Shiralkar, B S (1970) Two-phase flow and heat transfer in multirod geometries: a study of the liquid film in adiabatic air-water flow with and without obstacles. *General Electric, USA.* Report No GEAP-10248. [112,162,163,164]

Shiralkar, B S and Lahey, R T (1972) Diabatic local void fraction measurements in Freon-114 with a hot wire anemometer. *ANS Trans.* **15**, Part 2, 880. [125]

Shires, G L (1966) A comparison of two-phase venturi measurements at high and low pressures. *UKAEA* Report No SGHWR R & D Note 5. [96]

Shires, G L, Pickering, A R and Blacker, P T (1964) Film cooling of vertical fuel rods. *AEEW*-R343. [82,84]

Shires, G L and Riley, P J (1966) The measurement of radial voidage distribution in two-phase flow by isokinetic sampling. *AEEW*-M650. [118]

Shulman, H L and Molstad, M C (1950) Gas-bubble columns for gas-liquid contacting. *Ind. and Engng. Chem.* **42** (6), 1058-1070. [41]

Siboul, R (1976) Study of the significant output from screened thermocouples in the flow of water vapour which is not at equilibrium conditions. *CENG* Report No TT 514. [186]

Simmons, H C (1977) The correlation of drop-size distributions in fuel nozzle sprays. Part I: The drop-size/volume-fraction distribution. *J. Eng. Power.* **99** (3). Series A, 309-314. [152]

Simpson, H C and Brolls, E K (1974) Droplet deposition on a flat plate from an air-water mist in turbulent flow over the plate. *Symp. Multiphase Flow Systems, Univ. Strathclyde, April 1974.* Paper No A3. *I. Chem. E. Symp. Series*, No 38. [150]

Simpson, H C, Rooney, D H, Grattan, E et al (1977) Two-phase flow in large diameter horizontal tubes. *European Two-Phase Flow Group Mtg, Grenoble, June 1977.* Paper No A6. [26,108,185]

Simpson, O N, Hall, C, Rooney, D H and Hawlader, M N A (1977) Non-equilibrium effects in transient two-phase flow. *Conf. on Heat and Fluid Flow in Water Reactor Safety, Manchester, Sept 1977*, 19-25. [72]

Simpson, R L (1969) A study of adiabatic, fully-developed, annular-dispersed two-phase flow in a vertical round tube. *GEAP*-10094. [169]

Singer, R M and Holtz, R E (1970 Bubble growth measurements in non-uniformly superheated sodium. *Proc. ASME Winter Meeting, New York, Nov 1970.* 144-152. [77]

Singh, K, St Pierre, C C, Crago, W A and Moeck, E O (1969) Liquid film flow-rates in two-phase flow of steam and water at 1000 Lb/Sq. In. Abs. *AIChE Jnl.* **15** (1), 51-56. [171]

Sleath, J F A (1969) A device for velocity measurements in oscillatory boundary layers in water. *Journal of Physics E: Scientific Instruments*, **2**, Series 2, 446-448. [138]

Smirnov, O K, Zaitsev, V N and Serov, E E (1977) Investigation of burnout under transient hydrodynamic conditions. *Thermal Eng.* **24** (5), 72-74. [79]

Smith, A V (1971b) The transient denisty measurements in two-phase flows using X-rays. *J. Brit. Nucl. Energy Soc.* **10** (2), 99-106. [78]

Smith, A V (1975) Fast response multi-beam X-ray absorption technique for identifying phase distributions during steam-water blowdowns. *J. Br. Nucl. Ener. Soc.* **14** (3), 227-235. [51,109,121,182]

Smith, J E (1978) Gyroscopic/coriolis mass flow meter. *Canadian Controls and Instrummetation*, **29**. [101]

Smith, J W and King, D H (1975) Electro-chemical wall mass transfer in liquid particulate systems. *Can. J. Chem. Eng.* **53**, 41. [46]

Smith, R V (1971a) Two-phase, two-component critical flow in a venturi. *ASME* Paper No 71-FE-4. [70]

Smith, R V, Wergin, P C, Ferguson, J F and Jacobs, R B (1962) The use of a venturi tube as a quality meter. *Journal of Basic Engineering*, 411-412. [96]

Smith, W, Atkinson, G L and Hammitt, F G (1964) Void-fraction measurement in a cavitating venturi. *Jnl. Basic Engng.* 265-274. [49,70]

Solesio, J N, Flamand, J C and Delhaye, J M (1978) Liquid film thickness measurements by means of an X-ray absorption technique. *ASME Winter Annual Meeting, 1978.* [112]

Solomon, J V (1962) Construction of a two-phase flow regime transition detector. *M.Sc. thesis, Mass Inst. Tech.* [107]

Soo, S L , Trezek, G J and Dimick, R C (1964) Concentration and mass-flow distribution in a gas-solid suspension. *Industr. Engng. Chem. Fundam.* **3**, 98-106. [127]

Spigt, C L (1966) On the hydraulic characteristics of a boiling water channel with natural circulation. *Tech. U. Eindhoven, 1966.* [57,58,59]

Srinivasan, J and Lee, S L (1978) Measurement of turbulent dilute two-phase dispersed flow in a vertical rectangular channel by laser-Doppler anemometry. *ASME Winter Annual Meeting, San Francisco December 1978.* [156]

Stainthorp, F P and Wild, G J (1967) Film flow: The simultaneous measurement of wave amplitude and the local mean concentration of a transferable component. *Chem. Engng. Sci.* **22**, 701-704. [182]

Staniforth, R and Stevens, G F. (1965-66) Experimental studies of burn-out using Freon-12 at low pressure, with reference to burn-out in water at high pressure. *Proc. I. Mech. E.* **180** (30. [203]

Staniforth, R, Stevens, G F and Wood, R W (1965) An experimental investigation into the relationship between burn-out and film flow-rate in a uniformly heated tube. *AEEW*-R430. [171,203]

Staub, F W, Walmet, G and Neimi, R O (1968) Heat transfer and hydraulics. The effects of sub-cooled voids. *US-Euratom* Report No NYO-3679. [121]

Staub, F W and Zuber, N (1964) A program of two-phase flow investigation, 4th quarterly report, (Jan-March 1964). *General Electric*, Report No GEAP - 4577. [121]

Staub, F W, Zuber, N and Bijwaard, G (1967) Experimental investigation of the volumetric concentration in a boiling forced-flow system. *Nucl. Sci. and Engng.* **30**, 279-295. [78]

Stephens, A G (1977) Progress report on LOFT advanced densitometer. *US Nuclear Regulatory Commission, Proceedings of Meeting of Review Group on Two-Phase Flow Instrumentation, January 1977.* NUREG-0375 (Paper No I.5). [49].

Stephens, M J (1970) Investigation of flow in a concentric annulus with smooth outer wall and rough inner wall. Part 1: Transverse rib type roughness. *CEGB Report RD/BN 1535.* [161]

Sterlini, J and Trotignon, M (1971) A technique of local velocity and void fraction measurement in high velocity two-phase gas-liquid systems. *Symp. Two-Phase Flow, Israel, 1971.* Session 3. [128]

Stevens, G F, Elliott, D F and Wood, R W (1964) An experimental investigation into forced convection burn-out in Freon, with reference to burn-out in water. *AEEW*-R321. [203]

Stevens, G F, Elliott, D F and Wood, R W (1965) An experimental comparison between forced convection burn-out in Freon-12 flowing vertically upwards through uniformly and non-uniformly heated round tubes. *AEEW*-R426. [203].

Stevens, G F and Kirby, G J (1964) A quantitative comparison between burn-out data for water at 1000 Lb/In sq. and Freon-12 at 155 Lb/In sq. (abs) uniformly heated round tubes, vertical upflow. *AEEW*-R327. [204]

Stevens, G F and Macbeth, R V (1968) Further experimental confirmation of the Freon technique for scaling burnout. *European Two-Phase Flow Group Meeting, Oslo, June 1968.* Paper E3. [203]

Stevens, G F and Macbeth, R V (1970) Use of Freon-12 to model forced convection burn-out in water: the restriction on the size of the model. *Jnl. Brit. Nucl. Energy Soc.* **9** (4), 249-257. [205]

Stevens, G F and Macbeth, R V (1971) Use of Freon-12 to model forced convection burn-out in water: the restriction on the size of the model. *AEEW*-R683.

Stevens, G F and Wood, R W (1966) A comparison between burn-out data for 19-rod cluster test-sections cooled by Freon-12 at 155 Lb/In sq. (abs) and by water at 1000 Lb/In sq. in vertical upflow. *AEEW*-R468. [203,204,205]

Stevens, G F, Wood, R W and Pryzbylski, J (1968) An investigation into the effects of a cosine axial heat flux distribution on burnout in a 12ft long annulus using Freon-12. *AEEW*-R609. [203]

Stevenson, W H and Thompson, H D (1972) The use of the laser-Doppler velocimeter for flow measurements, Purdue Univ. *US Office of Naval Research Workshop, March 1972.* Proc. T73-05007 (2). [135]

Stiegel, G J, Shah, Y T (1977) Axial dispersion in a rectangular bubble column. *Can. J. Chem. Eng.* **55** (1), 3-8. [149]

Stothard, P H (1972) The measurement of two-phase flow parameters by salt tracer injection. *CEGB Report No CERL/RD/L/N 21/72.* [99]

Stothart, P H and Horrocks, J K (1972) A two-stream model of dispersion in annular two-phase flow. *CEGB Report No CERL/RD/L/N 18/72.* [149]

St Pierre, C C and Bankoff, S G (1967) Vapour volume profiles in developing two-phase flow. *Int. Jnl. Heat Mass Transfer.* **10**, 237-247. [49]

Strohmeier, W O (1974) Turbine flowmeters, past, present and future. *In* "Flow - its measurement and control in science and industry" 1, Part 2 (ISA), 687-693. [98]

Stuchly, S S, Rzepecka, M A and Hamid, M A K (1973) Microwave void-fraction monitor for organic coolants in nuclear reactors. *Microfiche* No AED-Conf, 409-001. [63]

Stukel, J J and Soo, S L (1968-69) Turbulent flow of a suspension into a channel. *Powder Technol.* **2**, 278-289. [118,127]

Styrikovich, M A, Baryshev, Y V, Tsiklauri, G V and Grigorieva, M E (1978) The mechanism of heat transfer between a water drop and a heated surface. *Proceedings of the 6th International Heat Transfer Conference, Toronto, Aug 1978*. **1**, 239. [83]

Styrikovich, M A and Nevstrueva, E I (1960) Investigation of vapour-content distribution in boiling boundary layer by the beta-radioscopy method. *Dokl. Akad. Nauk SSSR*, **130** (5), 1019-1022. [54]

Styrikovich, M A, Polonskii, V S and Bezrukov, E K (1971) A study of mass transfer in steam-generating channels by the 'salt method'. *High Temp.* **9** (3), 528-534. [149]

Subbotin, V I, Kirillov, P L, Smogalev, I P et al (1975) Measurement of some characteristics of a steam-water flow in a round tube at pressures of 70 and 100 atm. *ASME* Paper No 75-WA/HT-21. [114]

Subbotin, V I, Pokhvalov, Yu E and Leonov, V A (1977) The structure of steam-water slug flow. *Thermal Eng.* **24** (7), 48-51. [122,182]

Subbotin, V I, Polhvalov, Y E, Mikhailov, L E et al (1976) Time and structural characteristics of slug flow of gas/liquid mixture. *Thermal. Eng.* **23** (1), 48-51. [182]

Subbotin, V I, Sorikn, D N, Nigmatulin, B I, Milashenko, V I and Nikolayev, V E (1978) Integrated investigation into hydrdynamic characteristics of annular-dispersed steam-liquid flows. *Proceedings of the 6th International Heat Transfer Conference, Toronto, August 1978*. **1**, 327. [114,171]

Subbotin, V I, Sorokin, D N, Tzyganok, A A and Gribov, A A (1974a) Investigation of vapour bubbles effect on temperature of heat-transferring surface at nucleate boiling. *5th Internat. Ht. Transfer Conf. Sci. Council Japan, Sept 1974*. **4**. [192]

Subbotin, V I, Pokhvalov, Yu E, Mikhailov, L E, Leonov, V A and Kronin, I V (1974b) Resistance and capacitance methods of measuring steam contents. *Thermal Engng.* **21** (6), 88-99. [57]

Subramanian, G and Tien, C (1975) Induced liquid phase mixing due to bubble motion at low gas velocity. *Canadian J. Chem. Engng.* **53** (6), 611-620. [149]

Suezawa, Y, Matsumura, M and Nakajima, M (1972) Studies on cavitation erosion. *J. Basic Engng.* **94**, 521-532. [77]

Sullivan, J P and Ezekiel, S (1974) A two-component laser-Doppler velocimeter for periodic flow fields, *Journal of Physics E: Scientific Instruments*. **7**, 272-276. [137]

Sutey, A M and Knudsen, J G (1969) Comments on the application of the redox method of measuring mass transfer coefficients in two-phase (air-liquid) systems. *Int. Jnl. Heat Mass Transfer*, **12**, 373-374. [46,165]

Suzuki, S and Ueda, T (1977) Behaviour of liquid films and flooding in counter-current two-phase flow. Part 1: flow in circular tubes. *Int. J. Multiphase Flow*, **3**, 517. [191]

Swan, C L (1966) Photography of heated surface - fluid behaviour for transition boiling at 100 psia. *GEAP* 5094. [189]

Swanson, J L and Bowman, H F (1974) Transient surface temperature behaviour in nucleate pool-boiling nitrogen. *5th Internat. Ht. Transfer Conf. Sci. Council, Japan, Sept 1974*, **4**. [188]

Swartz, J E and Kessler, D P (1970) Single drop breakup in developing turbulent pipe flow. *AIChE J*. **16** (2), 254-260. [192]

Swithenbank, J, Beer, J N, Taylor, D S, Abbot, D and McCreath, G C (1976) A laser diagnostic for the measurement of droplet and particle size distribution. *Univ. Sheffield, Dept of Chemical Engineering and Fuel Technology report, 1976*. [155]

Tailby, S R and Portalski, S (1960) The hydrodynamics of liquid films flowing on a vertical surface. *Trans. Inst. Chem. Engrs*. **38**, 324-330. [177]

Takekoshi, T, Nishiwaki, I, Inami, K and Uchida, H (1969) Studies on the loss-of-coolant accident in water-cooled nuclear power reactors. *Crest Specialist Meeting, Depressurisation Effects in Water Cooled Reactors, Frankfurt, 1969*. [70]

Takeyama, T and Shimizu, S (1974) On the transition of dropwise-film condensation. Paper Cs 2.5. *5th Internat. Ht. Transfer Conf. Tokyo, Sept 1974*. 274-278. [188]

Taylor, D and Millican, B P (1974) Erection and testing of PFR secondary circuits and steam generators. *Proc. BNES. Conf. Fast Reactor Stations, March 1974*. 223-256. [86]

Technische Hogeschool, Eindhoven, Netherlands (1962) Heat Transfer and stability studies in boiling water reactor. Special technical report No 2. Study of possible application of acoustical methods for determining void fraction in boiling water reactors. *EURAEC* 278. [62]

Telles, A S and Dukler, A E (1970) Statistical characteristics of thin, vertical, wavy, liquid films. *Ind. Engng. Chem. Fundam*. **9**, 412-421. [112,177]

Termaat, K P (1970) Fluid flow measurements inside the reactor vessel of the 50 MWe dodewaard nuclear power plant by cross-correlation of the thermocouple signals, *Journal of Physics E: Scientific Instruments*, **3**, 589-593. [132]

Terry, F D, Kindred, R L and Anderson, S D (1965) Transient nuclear radiation effects on transducer devices and electrical cables. *Phillips Petrol Co*, IDO-17103. [76]

Thomas, D G, Baucum, W E, Bohanan, (1977) Quarterly progress report on blow down heat transfer, separate-effects programme, for Jan-March 1977. *ORNL*/NUREG/TM-109. [49]

Thomas, R L (1973) Vibration instrumentation for nuclear reactors. *Int. Symp. Vibration Problems in Industry, Keswick, April 1973*. Paper No 627. [86]

Thomas, R M (1976) Rules for modelling the steady-state carryunder performance of boiler drums using Freon-12. *CERL Report No CERL/RD/L/N22/76*. [203]

Thompson, B J (1974) Holographic particle-sizing techniques. *J. Phys. E: Sci. Instr*. **7**, 781-788. [152]

Thompson, D H (1968) A tracer-particle fluid velocity meter incorporating a laser. *Journal of Physics E: Scientific Instruments*, **1**,

Series 2, 929-932. [139]
Thwaites, G R, Kulov, N N and Nedderman, R M (1976) Liquid film properties in two-phase annular flow. *Chem. Eng. Sci.* 31 (6), 481-486. [26]
Tippets, F E (1964) Critical heat fluxes and flow patterns in high-pressure boiling water flows. *J. Heat Transfer*, **86**, 12-22. [189]
Toda, M, Ishikawa, T and Saito, S (1973) On the particle velocities in solid-liquid two-phase flow through straight pipes and bends. *J. Chem. Eng. Jap.* **6**, 140. [133]
Todreas, N (1974) Coolant mixing in LMFBR rod bundles. *USAEC Report No COO-2245-8*. [149]
Tokumitsu, M (1972) Full-scale safety experiments of FUGEN. *Crest Specialist Meeting on Emergency Core Cooling for Light Water Reactors, Munich*, MRR 115, **1**. [72]
Tolubinsky, V I, Kitchigin, A M and Povsten, S G (1974) Spectral analysis of noise in pool boiling in the transition region. *5th Internat. Ht. Transfer Conf. Sci. Council, Japan, Sept 1974.* **4**. [76]
Tolubinsky, V I and Konstanchuk, D M (1972) The rate of vapour-bubble growth in boiling of subcooled water. *Heat Transfer-Soviet Research*, **4** (6), 7-12. [192]
Tomida, T and Okazaki, T (1974) Statistical character of large disturbance waves in upward two-phase flow of air-water mixtures. *J. Chem. Engng. Japan, 1974.* **7** (5), 329-333. [112]
Tomida, T, Yoshida, M and Okazaki, T (1976) Liquid-side volumetric mass transfer coefficient in upward two-phase flow of air-liquid mixtures. *J. Chem. Eng. Japan.* **9** (6), 464-468. [41]
Tong, L S (1965) "Boiling heat transfer and two phase flow" John Wiley. [1]
Tong, L S and Hewitt, G F (1972) Overall viewpoint of flow boiling CHF mechanisms. *ASME paper 72-HT-54.* [65]
Torikai, K I, Hori, M and Akiyama, H (1964) Boiling heat transfer and burnout mechanism in boiling-water-cooled reactor. *3rd UN Conf. Peaceful Uses Atomic Energy*, Paper A/ONF.28/P/580. [192]
Torikai, K and Yamazaki, T (1965) The contact area of boiling bubbles on the heating surface. *Bull. of JSME*, **8** (32), 660-669. [192]
Torrance, J W (1973) Confusion and delusion in sonic flowmetering. *Advances in Instrumentation*, 28, Part 4 (ISA). [128]
Torrest, R S and Ranz, W E (1969) Improved conductivity system for measurement of turbulent concentration fluctuations. *I. and EC Fundam.* 8 (4), 810-816. [182]
Torrest, R S and Ranz, W E (1970) Concentration fluctuations and chemical conversion associated with mixing in some turbulent flows. *AIChE Journal.* 16 (6), 930-942. [182]
Towell, G D and Rothfield, L B (1966) Hydrodynamics of rivulet flow. *AIChE Journal*, **12**, 973. [174]
Traviss, D P and Rohsenow, W M (1973) Flow regimes in horizontal two-phase flow with condensation. *Proc. ASHRAE Spring Conf, May 1973.* **79** (II), 31-39. [106]
Trela, M (1974) Experimental investigations of the pressure drop of Freon-21 during flow boiling in vertical pipes. *Pol. Akad. Nank. PR. Inst. Masz, Przeplyw.* No 64, 59-78. [203]
Truong Quang Minh and Huyghe, J (1965) Measurement of average

thickness liquid film two-phase, two-component flow in the annular dispersed regime. *CEN Grenoble*, Report No 54. [114,190]
Tsiklauri, G V et al (1970) Determination of the rate of movement of the drops of a two-phase flow using the transit-time method. *High Temp.* **8** (4), 810-813. [139]
Tschoke, H and Moller, R (1977) Thermometry in fuel pin cladding tubes and fuel pin simulators. *Kerntechnische Gesselschaft im Deutschen Atomforum ev*, Proceedings of a Specialist Conference, Feb-March 1977, Hannover; "Experimental techniques in the field of thermodynamics and fluid dynamics. Part II; thermometry techniques and simulation of thermodynamics processes" 97, Interatom, Berg-Gladbach, West Germany. [80]
Turner, J T (1975) Wall pressure fluctuations associated with the flow of a bubbly air-water mixture in a vertical pipe. *U. New South Wales*. Report No 1975/FMT/1. [185]

Uchida, H (1969) The pressure pulse produced in water channel by rapid heating. *Crest Specialist Meeting, Depressurisation Effects in Water-Cooled Reactors, Frankfurt, 1969*. [77]
Uchida, H and Shimamune, H (1969) Test program of loss-of-coolant accident of water-cooled reactor in JAERI. *Crest Specialist Meeting, Depressurisation Effects in Water-Cooled React. Frankfurt, 1969*. [72,76]
Ueda, T and Tanaka, H (1975) Measurements of velocity temperature and velocity fluctuation distributions in falling liquid films. *Int. J. Multiphase Flow*, **2** (3), 261-272. [183,184]
Ueda, T, Tanaka, H and Koizumi, Y (1978) Dryout of liquid film in high quality R-113 upflow in a heated tube. *Proceedings of the 6th International Heat Transfer Conference, Toronto, August 1978*, **1**, 423. [150,153,169]
Uga, T, Mochizuki, I K and Yusa, H (1972) Measurement of the steam void-fraction in the Japan power demonstration reactor. (JPDR) *Hitachi-Review*, **15** (12), 468-474. [122]
Uhl, A E (1974) Tracer metering of compressible fluids. *In* "Flow - its measurements and control in science and industry" **1**, Part 2, (ISA), 809-821. [99]
Ulber, M and Lubbesmayer, D (1977) Acquisition and processing of temperature signals in blow down experiments. *Kerntechnische Gesellschaft im Deutschen Atomforum ev.* Proceedings of a Specialist Conference, Feb-March 1977, Hannover; "Experimental techniques in the field of thermodynamics and fluid dynamics. Part II; thermometry techniques and simulation of thermodynamics processes". 441, Interatom, Berg-Gladbach, West Germany. [72]
Upson, U L (1974) Evaluation of sensors for simulator annulus and related measurements in the plenum-fill experiment, Section I - Preliminary evaluation. *Battelle Pacific Northwest Laboratories*. [130]

Vander-Heyden, W H and Toschik, J F (1971) A new simplified nuclear magnetic resonance flowmeter, *Flow Symposium, Pittsburgh, PA, May 1971*. Paper No 2-7-180. [128]
Vander-Heyden, W H and Toschik, J F (1975) The nuclear magnetic resonance flowmeter: a new simplified design. *In* "Flow - its

measurement and control in science and industry", **1**, Part 2, 857-864. [128]

van der Molen and Galjee, F W B M (1977) Boundary layer and burnout phenomena in a subcooled two-phase flow. *European Two-Phase Flow Group Meeting, Grenoble, June 1977*. Paper No E1. [189]

van Paassen, C A A (1974) Thermal droplet-size measurements using a thermocouple. *Int. J. Heat Mass Transfer*. **17** (12), 1527-1548. [160]

van der Ros, T (1970) On two-phase flow exchange between interacting hydraulic channels. *Eindhoven U. of Technol. The Netherlands*. Report WW015-R160. [149]

van Stralen, S J D and Zijl, W (1978) Fundamental developments in bubble dynamics, *Proceedings of the 6th International Heat Transfer Conference, Toronto, August 1978*. **6**, 429. [192]

van Vonderen, A C M and van Vlaardingen, H F (1970) Impedance void gauge for cylindrical channels (inside cooling). *Europ. Two-Phase Flow Group Meeting, Milan, June 1970)*, Paper B10. [57,58]

vander-walle, F (1965) A study of the application of acoustical methods for determining void fractions in boiling water systems. *Symp. Two-Phase Flow. Exeter, 1965*. Paper E1. [62]

Varadi, G, Nabizadeh, H and Lording, N (1977) Development of new measurement techniques for determining phase velocity in a two-phase flow. *Kerntechnische Gesellschaft im Deutschen Atomforum ev* Proceedings of a Specialist Conference, Feb-March, 1977, Hannover; "Experimental techniques in the field of thermodynamics and fluid dynamics. Part II; thermometry techniques and simulation of thermodynamics processes" 387, Interatom, Berg-Gladbach, West Germany. [139]

Viecinz, H J and Mayinger, F (1977) Experimental and theoretical investigations in phase interaction and separation of a two phase mixture using the model fluid R12 (CF2CL2). *European Two-Phase Flow Group Mtg, Grenoble, June 1977*. Paper No B6. [203]

Vinokur, Ya G and Dil'man, V V (1959) Investigation of the bubble layer by the method of translucence of gamma rays. *Khim. Prom*. **7**, 719-621. [49]

Vogrin, J A (1963) An experimental investigation of two-phase, two-component flow in a horizontal converging-diverging nozzle. *Argonne Nat. Lab*. Report No ANL-6754. [49]

Vohr, J H (1963) A photography study of boiling flow. *USAEC*. NYO-9650. [189]

Votava, P (1974) Guide to pressure transducers, *Instruments and Control Systems*, 55-56. [27]

Wakstein, S C (1966) The motion of small particles suspended in turbulent air-flow in a vertical pipe. *Ph.D Thesis, Queen Mary College, London*. [132]

Waldram, K L, Fauske, H K and Bankoff, S G (1976) Impaction of volatile liquid droplets on to a hot liquid surface. *Can. J. Chem. Eng*. 54, 456. [83]

Walker, V and Rapier, A C (1965) Errors in the measurement of surface temperature caused by perturbations of the heat flux. *UKAEA* Report No TRG 1026(W). [34]

Wallis, G B (1962) General correlations for the rise velocity of cylindrical bubbles in vertical tubes. *General Electric*. Report No 62GL130. [12]

Wallis, G B (1969) "One dimensional two-phase flow" McGraw-Hill. [1, 7]

Wallis, G B, Steen, D A, Edgar, C B and Brenner, S N (1963) *Joint US-Euratom Research and Development Program, Quarterly Progress Report*, 10, 486. [107,169]

Walmet, G E and Staub, F W (1969) Pressure, temperature and void-fraction measurement in non-equilibrium two-phase flows. *Eleventh Nat. ASME/AIChE Ht. Transfer Conf. Minneapolis, Aug 1969*, 89-101. [51]

Walters, P T (1969) Optical methods for measuring water droplets in wet steam-flows. *CEGB* Report CERL RD/L/N-107/69. [154]

Walters, P T (1971) The optical measurement of water droplets in wet-steam-flows. *CEGB* Report No CERL/RD/L/R 1765. [154]

Walters, P T, Moore, K J and Langford, R W (1971) The design and performance of probes for making aerodynamic measurements in LP turbines. *CERL Lab. Report* No. RD/L/R 1724. [128]

Walton, A J (1963) Sonic methods for PFR channel blockage detection. *UKAEA, Risley,* TRG 665(R). [76]

Watson, G G (1964) A survey of techniques for measuring surface temperature. *Nat. Engng. Lab. Report No* 153. [34]

Waring, J P and Hochreiter, L E (1977) Flecht-set systems: Systems effects on reflooding behaviour. *Conf. on Heat and Fluid Flow in Water Reactor Safety, Manchester. Sept 1977*, 161-166. [82]

Wang, C P (1972) A unified analysis on laser-Doppler velocimeters. *Journal of Physics E: Scientific Instruments*. 5, 763-766. [135]

Wasserman, R and Grant, H (1973) Heated sensors for flow measurements. *Instrum. and Control Syst. USA*. 46 (5), 59-61. [183]

Webb, D R (1970a) Two-phase flow phenomena. *Ph.D Thesis, Cambridge Univ*. [28,30]

Webb, D R (1970b) Studies of the characteristics of downward annular two-phase flow. Parts 1-4. *AERE*-R6426. [28,30,112,113,177,185]

Webb, D R (1970c) Studies of the characteristics of downward annular two-phase flow. Part 5. Statistical measurements of wave properties. *AERE*-R6426, Pt 5. [112,113,177,178].

Webb, D R, Dukler, A E and Hewitt, G F (1970) Downwards cocurrent annular flow. *European Two-Phase Flow Group Meeting, Milan, June 1970*. Paper C1. [171]

Webster, J M (1971) A technique for the size and velocity analysis of high-velocity droplets and particles. *Brit. J. Photography*. (34). 752-757. [132,152]

Webster, J M, Weight, R P and Archenhold, E (1976) Holographic size analysis of burning sprays. *Combustion Flame*, 27 (3), 395-397. [152]

Weisman, J, Ake, T and Knott, R (1975) Two phase pressure drop across abrupt area changes in oscillatory flow. *U. Cincinnati*, Microfiche Report No COO-2152-18. [28]

Wenger, H C and Smetana, J (1972) Hydrogen density measurements using an open-ended microwave cavity. *IEE Trans. on Instrumentation and Measurement*. IM21, 2, 105. [63]

Wentz, L B, Neal, L G and Wright, R W (1968) X-ray measurement of void dynamics in boiling liquid-metals. *Nuclear Applications*, 4,

347-355. [78]
Wesley, R D (1977) Performance of drag-disc turbine and gamma densitometer in LOFT, *US Nuclear Regulatory Commission, Proceedings of Meeting of Review Group on Two-Phase Flow Instrumentation, January 1977*. NUREG-0375 (Paper No I.5). [93,98, 109].
Westinhouse Electric Corporation (1975) W/NRC FLECHT-SET and FLECHT program status. *Third Water Reactor Safety Research Information Meeting, Washington DC, Sept-Oct 1975*. [75]
Whalley, P B, Hutchinson, P and Hewitt, G F (1974) The calculation of critical heat flux in forced convective boiling. *5th Int. Ht. Transfer Conf. 1974. In* "Heat transfer 1974". **4**, 290-294, Scripta Book Co. [13,65,151]
Whalley, P B, Hutchinson, P and James, P W (1978) The calculation of critical heat flux in complex situations using an annular flow model. *Proceedings of the 6th International Heat Transfer Conf. Toronto, August 1978*. **5**, 65. [208]
Whalley, P B, Hewitt, G F and Hutchinson, P (1973) Experimental wave and entrainment measurements in vertical annular two-phase flow. *AERE*-R7521. [177,178,179]
Whalley, P B, Hutchinson, P and Hewitt, G F (1974) The calculation of critical heat flux in forced-convection boiling. *5th Internat. Heat Transfer Conf. Tokyo, Sept 1974*. Paper B6.11, 290-294. [13, 65,151]
Whalley, P B, Azzopardi, B J, Pshyk, L and Hewitt, G F (1977) Axial view photography of waves in annular two-phase flow. *European Two-Phase Flow Group Mtg, Grenoble. June 1977*. Paper No A5. [152, 190,191]
White, D F (1974) Velocity measurement by insertion meter. *Advances in Instrumentation*, **29**, Part 3 (ISA) [128]
White, E P and Duffey, R B (1975) A study of unsteady reflooding of water-reactor cores. *Trans. ANS*. **20**, 491. [82]
Wicks, M and Dukler, A E (1966) In-situ measurements of drop-size distribution in two-phase flow - a new method for electrically-conducting liquids. *Int. Mtg. Chicago*. [153]
Wigley, G (1977) The sizing of large drops by laser anemometry. *AERE*-R8771. [159]
Wigley, G and Hawkins, M G (1978) Three-dimensional velocity measurements by laser anemometry in a diesel engine cylinder under steady state inlet flow conditions and two different port geometries. *AERE*-R9088. [137]
Wikhammer, G A, Moeck, E O and MacDonald, I P L (1964) Measurement technqiues in two-phase flow, *Atomic Energy Canada Ltd*. Report No AECL-2215. [66]
Wilcox, S J and Rohsenow, W M (1969) Film condensation of liquid metals - precision of measurement. *Mass. Inst. Technol*. Report No DSR 71475-62. [35,36]
Wilkes, J O and Nedderman, R M (1962) The measurement of velocities in thin films of liquid. *Chem. Engng. Sci*. **17**, 177-187. [131]
Williams, D R and Stenhouse, I A (1978) Laser Raman gas diagnostic techniques. *AERE*-R9071. [145]
Williamson, K D (1970) Void fraction and related measurements in two-phase cryogenic flow systems. *Los Alamos Sci. Lab*. Report No

LA-DC-11238. [47]
Wilmshurst, T H (1971) Resolution of the laser fluid-flow velocimeter, *Journal of Physics E: Scientific Instruments*, **4**, 77-80.[137]
Winterton, R H S (1973) Boiling noise detection. *J. Brit. Nucl. Energy Soc.* **12** (4), 459-462. [76]
Woodmansee, D E and Hanratty, T J (1969) Base film over which roll waves propagate. *AIChE Journal*, **15**, 712-715. [191]
Woodward, D H (1964) Multiple light-scattering by spherical dielectric particles. *Jnl. Opt. Soc. Amer.* **54** (11), 1325-1331. [154]
Wurz, D (1971) Experimental study of the flow behaviour of a thin water-film and its reaction to a unidirectional air-stream, from moderate up to higher sonic velocity. *D. Eng. Thesis, Karlsruhe*, [128]
Wyler, J (1974) Practical aspects of hot-wire anemometry. *Instruments and Control Systems*, 57-60. [128]

Yablonik, R M and Khaimov, V A (1972) Determination of the velocity of inception of droplet entrainment in two-phase flow. *Fluid Mech. Soviet Res.* **1** (1), 130-134. [191]
Yamazaki, Y and Shibi, M (1968) Comparative study on the pressure drop of air-water and steam-water flows. *Int. Symp. Res. in Cocurrent Gas-Liquid Flow. U. Waterloo, Sept 1968.* **2**. [61]
Yau, A Y, Hamielec, A E and Johnson, A I (1968) Mass transfer with chemical reaction: oxidation of acetaldehyde in bubble reactors. *Int. Symp. Res. in Cocurrent Gas-Liquid Flow. Waterloo, Ontario, Sept 1968.* **2**. [41]
Ybarrondo, L (1975) Pressure measurement investigation - overview *Third Water Reactor Safety Research Information Meeting, Washington DC Sept-Oct 1975.* [109]
Ybarrondo, L (1975) Dynamic analysis of pressure transducer and two-phase flow instrumentation. *Third Water Reactor Safety Research Information Meeting, Washington DC Sept-Oct 1975.* [27,28]
Yih, T S and Griffith, P (1970) Unsteady momentum fluxes in two-phase flow and the vibration of nuclear system components. *Mass. Inst. Technol.* Report No DSR 70318-58. [87,89]
Yih, T S and Griffith, P (1970) Unsteady momentum fluxes in two-phase flow and the vibration of nuclear system components. *Argonne Nat. Lab.* Report No ANL-7685, 91-111. [87,89]
Young, R J, Yang, K T and Novotny, J L (1974) Vapour-liquid interaction in a high-velocity vapour jet condensing in a coaxial water-flow. *5th Internat. Ht. Transfer Conf. Tokyo, Sept 1974.* 226-230. [128]
Yu, S K W (1970) Nucleate boiling from groups of prepared cavities. *MSc Thesis, Univ. Manchester.* [192]

Zakharova, E A, Kolchugin, B A and Korniukhin, I P (1970) Vapour void-fraction in annular channels at separated and joint heat supplied. *4th Int. Heat Transfer Conf, Paris 1970*, **5**, Paper B 5.5. [49]
Zanelli, S and Hanratty, T J (1971) Relationship of entrainment to wave structure on a liquid film. *Int. Symp. Two-Phase Systems. Haifa, Israel, Aug-Sept 1971.* Session 2, Paper 1. [177]

Zanelli, S and Hanratty, T J (1973) Effect of entrainment on roll-waves in air-liquid flow. *Chem. Eng. Sci.* **28**, 643-644. [194]

Zenker, P (1972) Investigation into dust distribution in turbulently flowing dust-air mixtures in pipe lines. *Staub-Reinhalt Luft.* **32**, (1), 1-10. [117]

Ziegler, E E and Chiangkasri, I (1958) Gamma-ray techniques for the determination of the static volume quality of two-phase flow in a vertical pipe. *S B Thesis, MIT ME Dept.* [47]

Zielke, L A, Morgan, C D, Howard, C G and Currie, R L (1975) Rod bundle subchannel void fraction by gamma scattering. *Proceedings of the ANS Meeting - Reactor Fluid Flow and Heat Transfer, June 1975.* 412-413. [55,121]

Zirnig, W (1978) Beta absorption measuring system for steam-water-air mixture density in containment LOCKA experiments. *Paper presented at CNSI Specialist Meeting on Transient Two-Phase Flow, Paris, June 1978.* [54]

Zuber, N and Findlay, J A (1965) Average volumetric concentration in two phase flow systems. *J. Heat Transfer.* **87c**, 453. [8,12]

Zuber, N, Staub, F W, Bijwaard, G and Kroeger, P G (1967) Steady state and transient void fraction in two-phase flow systems. Final report for the programme of two-phase flow investigation. *GEAP-5417*, **1**. [28,51]

Subject Index

acoustics
   methods for void fraction, 62
   sound propagation in two-phase systems, 70-72
   ultrasonic detection of voids, 78
analytical models for two phase flow, 7-13
   one dimensional, 8-9
   momentum balance, 9-11
   homogeneous model, 10
   separated flow model, 11
   models based on flow patterns, 11-13
annular flow
   description of, 4-5
   mass-transfer coefficient in, 42
   void distribution in, 127
   mean film thickness in, 111-116
   laser anemometry in, 138
   droplet mass-transfer in, 146-149
   measurement of entrainment (film flow rate) in, 168-173
   axial view photography in, 190-191
applications
   of two phase systems, 2-3
autocorrelation function
   definition, 175
axial view photography
   in annular flow, 190-191
   stereoscopic, 191

balancing methods
   for wall shear-stress, 161-162
beta rays
   absorption, for void fraction, 54
   X-ray excitation, for void fraction, 55
blowdown (see system discharge) 69-76
   experiments on, 72-76
   problems in, 73-76
   dryout during, 78-82
bridge methods
   for critical heat-flux detection, 65-66
bubble flow
   description of, 4-5
   mass-transfer coefficient in, 42
   void distribution in, 120-127
   local velocity in, by double needle-probes, 133-135
   laser anemometry in, 138
   phase mixing in, 149
   bubble size-measurement in, 151-161
      photographic methods, 152
      light-scattering methods, 154-156
      from bubble velocity, 154-156
bubble growth and collapse, 76-78
   detection in reactors, 76-77
   in liquid sodium, 76-77
   photographic methods for, 77, 189
bubble size measurement, 151-161
   photographic measurement, 152
   light-scattering methods for, 154-156
   from bubble velocity measurement, 160
burnout, see critical heat flux

capacitance methods
   for void fraction, see impedance method
   for film thickness, 112
   for fluctuating film-thickness, (interfacial waves), 177

churn flow
  description of, 4
concentration distribution, 141-146
  sampling probes for, 141
  interferometric techniques for, 141-143
  laser holographic interferometry, 142-143
  Raman spectroscopy for, 143-145
concentration fluctuations (of both phases and components), 180-183
  void-fraction fluctuations, 180-182
    from X-ray absorption, 180-182
    using condutance needle-probes, 182
    using impedance meters, 182
    by neutron absorption, 182
    using capacitance probes, 182
    using beta absorption, 182
  concentration fluctuations, 182-183
    in liquid/liquid systems, 182
    in liquid films using two-colour light absorption technique, 182
    using fluorescence technique, 182-183
conductance (film) method
  for film thickness, 112-114
  for interfacial wave studies, 177-178
    power spectral-density analysis for frequency spectra, 178
    wave velocity from, by, cross-correlation, 178-179
  cross-correlation of output with wall pressure-fluctuations, 185
conductance (needle contact) devices
  in flow-pattern delineation, 107
  for film thickness measurement, 112-113
  for local void-fraction in bubble flow, 121-124
  contact hysteresis for, 122-123
  double needle device for local bubble-velocity measurement, 133
  double needle device for drop size, 153-154
  single needle device for drop size, 160
  in interfacial-wave studies, 177
  use in studying void fluctuations, 182
contact angle, 173-174
  by observation of triple interface, 174
  by reflection technique, 174
core melt-down
  vapour explosions in, 78
critical heat-flux (see dryout), 64-68
  detection by thermocouples, 64-67
    three-wire system, 66
    inside rod bundles, 66-67
    for indirectly-heated rods, 65-66
  bridge method for detection of, 65-66
  visual observation, 67
  infra-red detection system, 67
  under fault conditions, 78-80
critical two-phase flow, 69-70
  pressure-gradient measurement in, 70
cross correlation function
  definition, 176

dispersed flows
  laser anemometry in, 138
distribution
  of two-phase flows in manifold systems, 89-91
  of mass flow across channels, 116,120
  of void fractions, 120-128
  of velocity and momentum flux,

Subject Index

128-141
drag-disk devices, 130
drag screen
full flow, 100-101
drops
evaporation rate, 40
deposition rate of, 149-150
photographic studies of
coalescence of, 190-191
drop behaviour on surfaces
(Liedenfrost
condensation), 193
droplet deposition rate
see mixing characteristics
of two-phase flows
by film suction, 149-150
by sampling probe, 150
by fluorescence technique,
150
by magnesium oxide technique, 150
drop size measurement, 151-161
photographic methods, 152
holographic technique,
152
impingement (magnesium oxide)
coating technique for, 152-153
electrical contact methods
charge discharge method,
153-154
double probe (Wicks and
Dukler) method, 153
needle contact method,
160
light scattering methods
for, 154-156
thermal methods, 160
capillary collection method,
160
laser anemometry linked
techniques, 156-157
dryout (see critical heat-flux)
under fault conditions, 78-82
in blowdown, 78
with transient pressure
change, 79
with transient heat-flux,
79
with transient flow, 79
with simultaneous flow-and
pressure transients, 80

position of, in fault
conditions, 80

electrical discharge method for
droplet size, 153
electrochemical method
for mass-transfer coefficient,
44-47
for flow pattern via fluctuations in wall shear-stress,
107
for mean wall shear-stress,
164-165
characterisation of wall shear-stress fluctuations, 194-195
electrolytic methods (see
electrochemical method) for
tracer addition, 107
probe for velocity measurement,
135
electromagnetic flowmetering, 62-63
for void fraction, 62-63
entrainment (extent of, in
annular flow), 168-173
use of sampling and isokinetic
probes for, 169
use of film suction method for,
169-173
evaporating flows
phase mass-transfer in, 146-151
photographic studies of local
phenomena, in 189

fibre optic devices
for local void-fraction, using
contact detection, 125
for local void-fraction using
light-scattering, 125
film conductance method
for film thickness, 112-113
film flow rate (in annular flow)
168-173
from sampling and isokinetic-
probe measurement (by difference from total liquid flow),
169
by film suction
for total film-flow, 169-171
for local film-flow, 171-173
comparison of methods, 169
problems when flowrate high,
169

X-ray absorption for, 116
film suction method
  use in droplet mass-transfer studies, 150
  for measuring total film-flow, 169-171
  for local film-flow, 171-173
  comparison of methods, 169
  problems when flow rate high, 169
film thickness fluctuations (interfacial waves), 176-180
  photographic methods for, 176-177
  mirror-scanner device for, 177
  needle contact devices for, 177
  capacitance devices for, 177
  conductance (film) devices for, 177-179
    power spectral-density analysis of for frequency spectra, 177-179
    wave velocity by cross-correlation of signals from, 178-179
  light-absorption methods for, 178-179
  fluorescence technique for, 180
film thickness measurement, 111-116
  film conductance method, 112-113
  needle contact method for, 113-115
  fluorescence method for, 115-116
  reviews of,
  photographic measurements of, 112
  X-ray absorption technique for, 116
  capacitance method for, 112
first order parameters, 23-104
  definition of, 20-21
  primary design parameters (steady state), 23-68
  primary design parameters (fault conditions), 69-84
  secondary design parameters, 85-104

flow distribution
  in headers nad manifolds, 89-92
  see mass-flow distribution
flow pattern, 3-7, 105-111
  visual and photographic methods, 105-106
  X-radiography methods, 106
  needle contact (conductance probe) methods for, 107
  from fluctuation in wall shear stress (electrochemical), 107
  from fluctuation in X-ray absorption, 107
  pressure fluctuation analysis of for, 108-109
  multibeam X-ray and gamma absorption system for, 109-111
  from pressure fluctuations, 108-109
fluidised beds
  photographic studies of particle motion in, 189-190
fluorescence methods
  for film thickness, 115-116
  for particle deposition measurement, 150
  for film-thickness fluctuations (interfacial waves), 180
  for measurement of film-thickness and concentration fluctuations, 182
fluorinated hydrocarbons
  tables of properties, 198-202
fluorinated hydrocarbon modelling, 196-208
  applications, 196-204
  scaling laws for, 204-208
    empirical factors, 204-205
    dimensionless groups, 205-207
    predictive, 206-208
freon scaling
  see fluorinated hydrocarbon modelling
fuel-pin simulation
  in transients, 76

gas/solids flows
  heat-transfer in, 30-41
  isokinetic probes for, 118
  Pitot probes in, 128

Subject Index

tracer technique for mixing in, 149
gamma-ray absorption
  mechanisms of absorption, 47-48
  sources for, 50
  for void fraction, 47-54
    errors in, 51-54
    shaped collimators for, 53
    time-fluctuation effects, 53-54
  multi-beam system for flow pattern delineation, 109-111
  for void distribution, 121
gamma-ray scattering
  for void fraction, 55
  for void-fraction distribution, 121
gyroscopic/coriolis meter for mass flow, 101

heat-flux measurement
  in joule heating, 31
  in indirect heating, 31
  in radiant heat-transfer, 38
  from temperature fluctuation, 39
  in transients, 76
heat-flux meters, 38-39
heat transfer
  two phase, description of, 13-17
heat-transfer coefficient, 30-41
  typical measurement in two-phase flow, 31
  by transient response, 39
  mass-transfer analogy techniques for, 44-47
Hele-Shaw models
  for modelling two-phase flow in porous media, 197
holography
  for tracer particle observation (local velocity measurement), 132
  in interferometry, 132, 168, 188, 193
  for drop-size measurement, 152
homogeneous model
  for two phase flow, 10

hot-wire and hot-film anemometers
  in local void-fraction measurement, 125
  in single-phase or separated flows, 128
  for wall shear-stress, 162-164
  for velocity fluctuation measurement, 183-184

impedance method for void fraction, 57-60
  effect of flow pattern on, 57-58
  electrode design for, 58-59
  use in instability studies, 92
  use in studying void fluctuation, 182
infra-red detection
  of critical heat-flux, 67
  of temperature, 167
instability
  of two-phase boiling and condensing systems, 18-19, 91-92
instrumentation two-phase
  reviews of, 21-22
interferometry
  for concentration distribution, 141-143
    laser holographic, 141-143
  for temperature distribution, 168
    laser holographic, 168
  for temperature-fluctuation observation, 188
    laser holographic, 188
  in photographic studies, 193
    laser holographic, 193
isokinetic probe
  for mass-flow measurement, 117-120
  for local void-fraction measurement, 127
  combined with momentum-flux measurement, 129-130
  use in entrainment (film flow) measurement, 168
  throat collection device, 120
impact temperature
  on thermocouples, 166
interfacial waves (film thickness

fluctuation, 176-180
  photographic methods for, 176
  mirror scanner device for, 177
  needle contact devices for, 177
  capacitance devices for, 177
  conductance (film) device for, 177-179
    power spectral density, analysis of, for frequency spectra, 177-179
    wave velocity by cross-correlation of signals from, 178-179
  light-absorption method for, 179-180
  fluorescence technique for, 180
  photographic studies of, 190-191
    axial view photography for, 190-191
infra-red absorption
  for void fraction, 64

laser anemometry, 135-138
  alternative methods for, 135-136
    Doppler frequency shift, 135-136
    fringe method, 136-137
  application in dispersed flows, 138
  application in film flows, 138
  application in bubble flow, 138
  limitations in complex flows, 138
  for velocity fluctuation, 184-185
  for drop size, 156-160
laser holographic interferometry, 142-143
laser Raman spectroscopy (see Raman spectroscopy)
light-absorption method
  for studying film-thickness fluctuations, 178-180
  two colour method for concentration fluctuations, 182
light-scattering techniques
  for void distribution, 127
  for velocity measurement
    laser anemometry, 135-138
    cross-correlation method, 139
  for bubble and drop size measurement, 154-156
liquid-level detection
  impedance devices for, 103
  electric pulse reflection techniques for, 103-104
local phenomena, photographic observation of, 189-194
  reviews of, 189
  refraction effect causing problems in, 189
  in adiabatic flows, 189
  in evaporating flows, 189
  in fluidised beds, 189-190
  in interfacial-wave studies, 190-191
    axial view photography for, 190-191
  in bubble and drop coalescence and break-up studies, 191-192
  in nucleate pool-boiling, 192
  in studies of drop behaviour and growth on surfaces, 193
  interferometric methods for, 193
  schleiren methods for, 193
  shadowgraph methods, 194

magnesium oxide coating technique
  for droplet deposition studies, 150
  for drop-size measurement, 152-153
manifolds
  two-phase flow distribution, 89-91
mass flow distribution, 116-120
  sampling probes for, 116-117
  isokinetic probes for, 117-119
  wall scoop-probe for mass-flow near wall, 118-119
mass-flow measurement
  in transients, 73
  in steady state, 92-103
    orifice plate for, 96

Subject Index

venturimeters for, 96-97
turbine meters for, 98-99
by tracer injection, 99-101
pulsed neutron activation techniques for, 100
true mass flow meter, 101-102
    gyroscopic-coriolis, 101
    turbine, 101-102
mass transfer coefficients, 41-47
in bubble flow, 41-42
in slug flow, 42
in annular flow, 42-43
napthalene deposition technique for, 44
rubber coating method for, 44-45
electrochemical method, 44-47
mass transfer, of phase, see mixing characteristics
see phase mass-transfer
mean phase content (void fraction), 47-64
microthermocouples, 185-188
use in measuring temperature, fluctuations, 185-188
    combined needle contact and microthermocouple, 187-188
    use in boiling, 186
    use in flashing flow, 187
momentum flux
in blowdown, 75
fluctuating, causing vibrations, 85-89
local values, by combined isokinetic/momentum probes, 129-130
drag-disk devices for, 130-131
momentum vector sensor, 130
distribution of, 128-141
microwave absorption, for void fraction, 63
mirror-scanner device
for interfacial wave velocity measurements, 177
mixing characteristics of two-phase flows, 146-151
tracer techniques for, 146-149
for droplet mass-transfer
in annular flow, 146-149
in bubble flows, 149
in gas/solid flows, 149
inter-subchannel mixing studies using, 149
see phase mass-transfer
multibeam gamma absorption method
in flow-pattern studies, 109-111
for void distribution, 121

naphthalene deposition
for mass transfer coefficient measurement, 44
needle contact (conductance) devices in flow-pattern delineation, 107
for film thickness measurement, 113-115
for local void-fraction in bubble f-ow, 121-125
contact hysteresis for, 122-123
double-needle device for local bubble velocity measurement, 133-134
double-needle device for drop size, 160
in interfacial wave studies, 177
use in studying void fluctuation, 182
neutron absorption
for mean void-fraction, 54
for studying void fluctuation, 182
neutron scattering
for void fraction, 55
non-equilibrium vapour-liquid mixture
possible application of Peltier effect thermocouples to temperature measurement in, 167
nuclear magnetic-spin resonance
for void-fraction measurement, 63
for velocity measurement, 128
nucleate boiling
detection of initiation of, 76-77

fluid-temperature fluctuation in, 186-187
wall-temperature fluctuation in, 188
photographic studies of, 192-193

optical methods
  for void fractions, 63
  for void distribution, 125-126
  for contact angle, 174
optical probes
  for local void fractions, 125-126
orifice plate
  for transient mass flow measurement, 73
  for mass flow/quality measurement, 96-97
optical pyrameters
  temperature measurement, 34

particle size measurement, 151-161
  see-drop-size measurement
  see bubble size measurement
Peltier effect thermocouples
  possible use for temperature distribution in non-equilibrium flows, 167
phase concentration distribution (local void fraction), 120-128
  by gamma absorption, 121
  by gamma scattering, 121
  multi-beam gamma method for, 121
phase mass transfer (mixing), 146-151
  see mixing characteristics
  droplet deposition measurement
    film suction technique, 150
    by sampling probe, 150
    by magnesium oxide coating technique, 150
    by fluorescent particle method, 150
    by measuring burnout flux, 150-151

photography
  in flow-pattern studies, 105-106
  for local velocity measurement, 130
    with tracers, 131
  for drop-and bubble-size measurement, 184
  see local velocity fluctuation measurement, 184
  see local phenomena, photographic observation of, 189-194
  reviews of, 189
  refraction effect, 189
  axial view, 190-191
  interferometric methods for, 193
  schleiren methods for, 193
  shadow-graph methods for, 194
photolysis, flash (for tracer addition), 133
plug flow
  description of, 4-5
Pitot probes, 128-129
  special collecting device, 120
pressure, 24-30
  pressure transducers, 26-30
  in transients, 75
power spectral density
  definition, 176
pressure drop, 24-30
  problems in measuring, 25-26
  keeping lines single phase, 26
  liquid purging, 26-27
  DP cells for, 28-30
  manometers for, 24-26
  pressure transducers for, 26-29
  in critical flow, 70
pressure fluctuations, 185
  use in void-fraction measurement, 63-64.
  wall pressure fluctuations, 185
    correlation with film-thickness fluctuations, 185
  for flow pattern characteristics, 108-109
pressure pulses
  transmission, 71-72
  generated by bubble growth and collapse, 77-78

Subject Index

pressure transucers
  types of, 29
  reviews of, 27
Preston tubes, 165
primary design parameters, fault conditions, 69-84
  definition of, 20-21
primary design parameters, steady state, 23-68
  definition of, 20
pulse neutron activation technique
  for mass flow measurement, 100

quality
  measurement of, 92-103
  by measurement of momentum flux, 93
  orifice plate for, 96-97
  venturis for, 96-97
  turbine meters for, 98-99
  by tracer methods, 99-100
quick-closing valve technique for void fraction, 61-62
  application to heated systems, 61-62
  in transients, 61-62

radiation, nuclear
  effect on instrumentation, 76
radiation, thermal effect on thermocouples, 106
radioactive tracer methods, see tracer methods
Raman spectroscopy
  concentration measurement by, 143-145
  temperature measurement by, 143-145
reflooding (see rewetting), 82-84
resistance bulbs
  for temperature measurement, 32
reviews
  two-phase instrumentation, 22
  pressure transducers, 27
  temperature measurement, 31-32
  critical heat-flux mechanisms, 64-65
  of rewetting, 83
  of film thickness measurement, 112
  of photographic methods, 189
rewetting, 82-84
  caused by thermocouples, 80
  types of, 82
  review of, 83
  promotion of by spacer grids, 83
  effect of multi-component systems in, 83-84
  effect of surface tension gradients on, 83-84
  surface condition effect on, 84
  coupled effect of velocity and subcooling on, 84
rupture mechanism
  for blowdown experiments, 75

sampling probes
  for mass-flow distribution, 116-117
  for concentration distribution, 141
  use in droplet mass-transfer studies, 150
  use in entrainment (film flow) measurement, 169
scaling of two-phase system, 196-208
  fluorinated hydrocarbon scaling, 196-208
  Hele-Shaw model for porous media, 196
  in liquid steel processes, 196
schleiren methods
  in photography of local phenomena, 193
scoop probe
  for mass-flow measurement near wall, 118-119
secondary design parameters, 85-104
  definition of, 21
second order parameters, 105-174
  definition of, 21
separated flow model, 11
shear stress, see wall shear stress
shadow-graph methods

in photography of local
phenomena, 194
shock wave
transmission, 71-72
slug flow
description of, 4-5
mass-transfer coefficient
in, 42
sound wave
propagation in two phase
flow, 70-71
spray cooling (see rewetting),
82
Stanton tube, 165
stability
of two-phase boiling and
condensing systems, 91-92
strain gauge flowmeters, 138-
139
stratified flow
description of, 5
subcooled boiling
temperature distribution
in, 168
surface tension
gradients of, effect on
wetting, 83-84
system discharge and related
parameters, 69-76
   see blowdown
subchannels
tracer techniques for
mixing between, 149

T-junction
use in measuring momentum
flux, 87-89
two-phase flow division at,
89-91
taumeter, 161-162
temperature distribution, 165-
168
in subcooled boiling, 168
in non-equilibrium vapour-
liquid mixture, possible
use of Peltier effect for,
167
temperature measuring
system for
thermocouples, 167
platinum resistance
thermometers, 167
thermal noise measure-
ment, 167
infra-red, 167
using interferometry, 167
with holography, 167
temperature fluctuations, 185-189
use in heat-flux measurement,
39-40
microthermocouples for measuring in fluids, 185-189
combined needle-contact
device and microthermo-
couple, 186-187
use in boiling, 186
use in flashing flow, 187
interferometric methods for,
in fluids, 188
in nucleate boiling, 188
with holography, 188
wall-temperature fluctuation,
188
in nucleate boiling, 188
in dropwise condensation,
188
surface resistance ther-
mometer for, 188
temperature measurement
of wall, 34-36
reviews of, 34
during transients, 80
in temperature distribution
studies, 166-167
resistance bulbs for, 32
thermistors for, 32
fluid filled thermometers for,
34
optical pyrometers for, 34
bimetallic thermometers for,
34
by Raman spectroscopy, 143-145
thermal methods
for drop-size measurement, 160
thermal tracers
for local velocity measurement,
132-133
thermal noise for temperature
measurement, 167
thermistors
for temperature measurement,
32
thermocouples
list of main types, 31
response of, 32
in wall temperature measure-

Subject Index

ment, 34-36
three wire systems, 66
in critical heat-flux
detection, 66-67
in temperature distribution
measurement, 167
thermometers
impact temperature effect, 166
radiation effects, 166
wetting effect, 160
fluid filled, 34
bimetallic, 34
third order parameters, 175-195
definition of, 21
tracer methods
in mass flow/quality
measurement, 99-100
in local velocity measurement
using tracer particles, 130-132
electrolytic tracer addition, 132
thermal tracers, 132-133
flash photolysis, 133
radioactive tracers, 133
in mixing and phase mass-transfer studies, 146-149
for droplet mass transfer in annular flow, 146-149
in bubble flows, 149
in gas/solids flows, 149
in inter-sub-channel mixing, 149-150
particle tracers in velocity fluctuation measurement, 185
transients
description of, 18-19
mass-flow measurement in, 73
void fraction measurement in, 73-75
dryout during, 78-82
turbine meters
for quality/mass flow measurement, 98-99
transfer response of, 98-99
for true mass-flow, 101-102
two phase flow and heat transfer, introduction to, 1-19
application of, 2-3
physical nature of, 3-7

flow patterns in, 3-7
analytical models for, 7-13
one dimensional, 7-8
two dimensional, 8-9
momentum balance, 9-11
homogeneous model, 10
separated flow model, 11
models based on flow patterns, 11-13
two phase heat transfer description, 13-17
unstable and transient flows description, 18-19
two-phase instrumentation
reviews of, 22

ultrasonic detection of voids, 78
ultrasonic velocity measurement, 128

vapour explosions
in core melt down, 78
velocity distribution, 128-141
in separated flows, 128-129
hot wire for, 128
ultrasonic flow-metering, 128
nuclear magnetic resonance, 128
vortex shedding flowmeters, 128
multihole pitot probes, 128
use of Pitot probes in, 128-129
use of isokinetic probes in, 129-130
combined momentum flux measurements with, 129-13)
drag disk devices, 130-131
momentum vector sensor for, 130
photographic technique for, 131-132
tracer methods for, 131-134
using tracer particles, 131-132
electrolytic tracer addition, 132
thermal tracers, 132-133
flash photolysis, 133
radioactive tracers, 133

needle contact method for, 133
electrolytic probe, for, 135
laser anemometry, 135-318
   alternative methods, 135-137
   application in dispersed flows, 138
   application in film flows, 138
   limitation in complex flows, 138
strain gauge method, 138-139
light scattering method, 139-140
double optical probes for, 133-135
velocity fluctuation, 183-185
hot film/wire sensors for, 183-18
photographic measurement of
   use of tracer particles in, 184
laser anemometry in, 184-185
venturi meters
   use in quality measurement, 96-97
vibration
   due to fluctuation in two-phase momentum flux, 85-89
   techniques for measurement, 86
   mechanisms in cross flow, 85
visual observation
   of critical heat flux, 67
void detection
   in reactors (bubble generation), 76-77
void fraction distribution, 120-128
   by gamma absorption, 121
   by gamma scattering, 121
   multi beam gamma method form, 121
   using needle contact probes, 121-125
   using hot wire and hot film anemometers, 125
   using optical probes, 125-126
   two-wire conductance probe for, 127
   using light scattering, 127
   using isokinetic probes, 127-130
void fraction, fluctuating
   impedance measurements of, for detection of instability, 92
   dilatometer measurement of in instability studies, 92
   from X-ray absorption, 180-182
   measurement using conductance needle probes, 182
   using impedance meters, 182
   by neutron absorption, 182
void fraction (mean), 47-64
   gamma ray absorption, 47-54
   X-ray absorption, 47-54
   β ray absorption, 54
   neutron absorption, 54
   neutron scattering, 55
   X-ray excitation by β-rays, 55
   neutron excitation with γ rays, 55
   gamma scattering, 55
   impedance methods, 57-60
   quick closing valve method, 61-62
   by total volume measurement, 61
   acoustic methods, 62
   by phase velocity measurement, 62
   nuclear magnetic resonance, 63
   infra-red absorption, 64
   electromagnetic flowmeter, 62-63
   optical methods for, 63
   microwave absorption, 63
   from pressure (flow fluctuations), 63-64
   in transients, 73-75
   nuclear magnetic resonance, 63
void growth
   X-ray methods for, 78
   ultrasonic detection of, 78
vortex flow-meter, 128

wall shear-stress

fluctuations in, for flow-
pattern delineation, 107
balancing methods for, 161-
162
    tau (τ) meter, 161-162
heat and mass-transfer
methods for, 162-165
    electrochemical method,
    107, 164-165
    hot-film sensors, 162-
    164
impact probe methods, 165
    Preston or Stanton tube
    for, 165
characterisation of fluctuations in, 194-195
    electrochemical methods
    for, 194-195
wall-temperature measurement,
34-36
    in electrically-heated
    systems, 34-36
    in fluid-heated systems,
    34-36
    by phosphorescence, 37
    with fluctuations, 188
        in nuclear boiling, 188

        in dropwise condensation,
        188
    surface resistance ther-
    mometers for, 188
waves, see interfacial waves
wavy flow
    description of, 5
wispy annular flow
    description of, 4

X-radiography
    for flow pattern studies,
    106
X-ray absorption
    for void fraction, 51-54
        errors in, 51-54
    for void growth in transient,
    78
    use of fluctuations in, for
    flow pattern delineation,
    107-108
    multibeam for use in flow
    pattern studies in horizontal
    tubes, 109-111
    for void fluctuations
    characterisation, 180-181
    for film thickness, 116